Respiratory Muscles:
Structure, Function & Regulation

Colloquium Series on Integrated Systems Physiology: From Molecule to Function to Disease

Editors

D. Neil Granger, *Louisiana State University Health Sciences Center–Shreveport*

Joey P. Granger, *University of Mississippi Medical Center*

Physiology is a scientific discipline devoted to understanding the functions of the body. It addresses function at multiple levels, including molecular, cellular, organ, and system. An appreciation of the processes that occur at each level is necessary to understand function in health and the dysfunction associated with disease. Homeostasis and integration are fundamental principles of physiology that account for the relative constancy of organ processes and bodily function even in the face of substantial environmental changes. This constancy results from integrative, cooperative interactions of chemical and electrical signaling processes within and between cells, organs, and systems. This eBook series on the broad field of physiology covers the major organ systems from an integrative perspective that addresses the molecular and cellular processes that contribute to homeostasis. Material on pathophysiology is also included throughout the eBooks. The state-of the-art treatises were produced by leading experts in the field of physiology. Each eBook includes stand-alone information and is intended to be of value to students, scientists, and clinicians in the biomedical sciences. Since physiological concepts are an ever-changing work-in-progress, each contributor will have the opportunity to make periodic updates of the covered material.

Published titles

(for future titles please see the Web site, www.morganclaypool.com/page/lifesci)

Respiratory Muscles: Structure, Function & Regulation
Gary C. Sieck and Heather M. Gransee
www.morganclaypool.com

ISBN: 9781615043842   paperback

ISBN: 9781615043859   ebook

DOI: 10.4199/C00057ED1V01Y2012ISP034

A Publication in the

COLLOQUIUM SERIES ON INTEGRATED SYSTEMS PHYSIOLOGY: FROM MOLECULE TO FUNCTION TO DISEASE

Lecture #34

Series Editors: D. Neil Granger, LSU Health Sciences Center and Joey P. Granger, University of Mississippi Medical Center

Series ISSN
ISSN 2154-560X   print
ISSN 2154-5626   electronic

Respiratory Muscles:
Structure, Function & Regulation

Gary C. Sieck and Heather M. Gransee
Mayo Clinic
Department of Physiology & Biomedical Engineering

COLLOQUIUM SERIES ON INTEGRATED SYSTEMS PHYSIOLOGY:
FROM MOLECULE TO FUNCTION TO DISEASE #34

 MORGAN & CLAYPOOL LIFE SCIENCES

ABSTRACT

Breathing is usually automatic and without conscious effort; yet our breathing is a complex motor function requiring the coordinated activation of a number of respiratory muscles that span from our heads to our abdomen. Some of our respiratory muscles serve to pump air into and out of our lungs (ventilation). These pump muscles act on the thoracic and abdominal walls and are all skeletal muscles. Other respiratory muscles in our bodies control the caliber of the passageway for air to enter our lungs. These airway muscles include skeletal muscles of the head (e.g., tongue and suprahyoid muscles) and neck (infrahyoid, pharyngeal and laryngeal muscles), as well as smooth muscles that line our trachea and bronchi down to the alveoli where gas exchange occurs. This book provides an overview of the anatomy and physiology of our respiratory muscles, including their neural control. This book also includes an overview of the basic structure and function of both skeletal and smooth muscles. The two basic types of respiratory muscles (skeletal and smooth muscle) vary considerably in the organization of their contractile proteins and the underlying mechanisms that lead to force generation and contraction, including their neural control.

KEYWORDS

breathing, skeletal muscle, smooth muscle, diaphragm muscle, cross-bridge, anatomy, physiology, sarcomere, neural control, contractile proteins, motor units, recruitment

Contents

CHAPTER 1

Introduction

"Until I feared I would lose it, I never loved to read. One does not love breathing."
(Harper Lee—American author)

"What can we do but keep on breathing in and out, modest and willing, and in our places"
(Mary Oliver—American poet)

Breathe in, breathe out—the act of breathing comes naturally to us, without conscious effort. Yet the control of our breathing is quite complex involving the coordinated activation of respiratory skeletal muscles of the head, neck, chest wall and abdomen, and smooth muscles lining the trachea and airways of our lungs. As recognized in the quotes above, we take breathing for granted, but become aware of our breathing only when it becomes labored due to dysfunction of the respiratory muscles. This book will provide an overview of the physiology (and in some cases, the pathophysiology) of our respiratory muscles, arguably the most important muscles in our body.

Respiratory muscles serve to move air into and out of our lungs (ventilation), thereby affecting pulmonary gas exchange that supplies O_2 to our arterial blood and eventually the metabolically active tissues of our bodies while removing CO_2 from the blood. The muscles involved in ventilation, the actual movement of air into our lungs, are called pump muscles, and they are all skeletal (or striated) muscles. There are also other types of respiratory muscles that control the caliber of the conductive airways to our lungs both above and below the larynx. Some of these airway muscles are also skeletal muscles, and these fall into two categories, 1) those muscles that control the patency of the nasal, oral, and pharyngeal conductive pathway for air, termed upper airway muscles and 2) muscles that control the opening (abductors) or closing (adductor) of the laryngeal inlet. Smooth muscles also line our airways from the trachea and bronchi down to the alveoli where gas exchange occurs. Contraction and relaxation of these airway smooth muscles control the patency of our airways increasing or decreasing resistance to airflow.

To understand the physiology of our respiratory muscles, it is important to understand the basic structure and function of both striated and smooth muscles. These two types of muscle vary considerably in the organization of their contractile proteins and the underlying mechanisms that lead to force generation and contraction, including neural control and excitation–contraction coupling.

. . . . .

CHAPTER 2

Respiratory Pump Muscles

The pump muscles are skeletal muscles that serve to move air into (inspiratory) and out of (expiratory) our lungs. Thus, the pump muscles are broadly categorized as inspiratory or expiratory muscles based on their mechanical action on the chest wall and thoracic cavity that translates into a decrease (inspiratory) or increase (expiratory) in thoracic pressure.

2.1   DIAPHRAGM MUSCLE

The diaphragm muscle is unique to mammals, and it is the major muscle involved in inspiration. The diaphragm muscle separates the thoracic (pleural) and abdominal (peritoneal) compartments of our bodies. With activation and contraction, our diaphragm muscle moves caudally, causing an increase in the vertical dimension of the thoracic cavity. This creates a negative thoracic pressure, which is transmitted to our lungs via the pleura that line both the chest wall and the lungs. Thus, with diaphragm muscle contraction and generation of a negative intrathoracic pressure, our lungs inflate with air inspiration. As our diaphragm muscle contracts and moves caudally, our abdominal cavity is compressed and abdominal pressure increases. Accordingly, with diaphragm muscle contraction, there is a marked increase in the pressure difference between our thoracic and abdominal cavities (an increase in transdiaphragmatic pressure). As our diaphragm muscle relaxes, this transdiaphragmatic pressure difference rapidly decreases, due in large part to the passive recoil forces of our lungs and chest wall. This rapid decrease in transdiaphragmatic pressure causes air to move out of our lungs, a process known as expiration. Thus, the diaphragm muscle is an active inspiratory pump for lung ventilation. If activation of our pump muscles fails to generate sufficient intrathoracic pressure then ventilation of our lungs cannot be sustained.

Diaphragm muscle fibers originate from the entire circumference of our thoracic cavity and insert into a central tendon. Based on the origin of muscle fibers, the diaphragm muscle has been separated into three major regions. Muscle fibers originating from the xyphisternal junction (the joint connecting the xyphoid and sternum) form the sternal region of our diaphragm muscle. Diaphragm fibers that originate from the broad expanse of our lower rib cage form the costal region of our diaphragm muscle. A clearly separate group of diaphragm muscle fibers originate from our

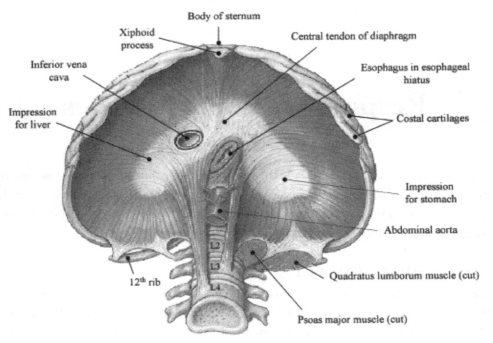

FIGURE 1: Inferior view of the diaphragm, specifically illustrating the origin and insertion of muscle fibers. Used with permission from DiMarco [23].

upper lumbar vertebrae and form the crural region. Fibers from each of these diaphragm muscle regions insert into the central tendon (Figure 1).

Our diaphragm muscle is really a paired muscle comprising right and left sides that are generally symmetrical. The orientation of fibers in the two sternal regions of our diaphragm muscle is essentially parallel, both within each side and across the left and right sides.

Muscle fibers in the costal region of each side radiate inward from the much more expansive circumference of the costal margin of our lower rib cage to the smaller circumference of the central tendon. Thus, the orientation of fibers in each of the two costal regions of our diaphragm muscle is not parallel. The direction of force vectors generated in each side of the costal region of the diaphragm muscle pulls in roughly opposite directions. Moreover, the orientation of fibers in the costal regions of our diaphragm muscle is not flat but curved in a dome shape; thus increasing the overall length of these fibers. Contraction or shortening of muscle fibers in the costal regions causes the curvature of the diaphragm muscle to flatten downward, thereby pushing on the abdominal cavity and increasing abdominal pressure.

The orientation of diaphragm muscle fibers in the crural region is even more complex and varies somewhat on the right and left sides. The right portion of the crural diaphragm is larger (a greater number of muscle fibers versus large muscle fibers) and longer than the left. Muscle fibers in the right crural diaphragm arise from the anterior surfaces of the bodies and intervertebral fibrocartilages of our upper three lumbar vertebrae. In the left crural region, diaphragm muscle fibers originate from the corresponding parts of the upper two lumbar vertebrae only. The medial margins of the right and left crural regions project anteriorly and medially to meet in the midline forming an arch across the front of our descending aorta. Thus, as it passes from the thoracic to the abdominal cavity, the descending aorta is dorsal to the crural regions of the diaphragm muscle. This anatomical relationship is quite important since this means that inspiratory contraction of our diaphragm muscle fibers will not impede blood flow in our descending aorta. Similarly, it would not be advantageous for contraction of our diaphragm muscle to restrict blood flow in the inferior vena cava as it traverses from the abdominal cavity to thoracic cavity carrying venous blood back to the heart. In fact, there is an advantage for the increased transdiaphragmatic pressure during inspiration to promote an increase in venous return of blood to the right atrium, especially when we are standing since in this case, the return of venous blood to the heart must work against gravity. Thus, our inferior vena cava passes through the central tendon of the diaphragm where muscle fiber contraction will not constrict blood flow.

On our left side, the esophagus passes through the crural region of the diaphragm muscle. Thus, contraction of muscle fibers in the left crural region of the diaphragm acts as a sphincter during inspiration. Physiologically, this is important since during inspiration, the increase in transdiaphragmatic pressure would tend to push gastric contents from our stomach, located in the abdominal cavity, into our esophagus, located in the thoracic cavity, unless contraction of fibers of the left crural region closed off the esophagus.

Muscle fibers in each of the regions of our diaphragm insert into the central tendon, which is a thin but strong aponeurosis. The central tendon of the diaphragm is situated somewhat closer to the anterior chest wall such that diaphragm muscle fibers originating from the posterior costal margin are slightly longer. Rostrally, the central tendon is situated immediately below the pericardium, with which it partially blends. The pericardium is a reflection of the pleura, which also covers the rostral surface of our diaphragm muscle. Caudally, the central tendon is situated above the liver and stomach. A portion of the liver connects to the diaphragm directly (the so-called bare area), but most of the liver is covered by the visceral peritoneum, which is a thin, double-layered membrane similar to the pleura that also serves to reduce friction against other organs. There are reflections or folds of peritoneum that appear as ligaments and maintain the position of the relatively heavy liver in the abdominal cavity. These folds or ligaments include the falciform, right and left triangular, coronary and round ligaments of the liver. The inferior surface of our diaphragm muscle is covered by the peritoneum, and these peritoneal "ligaments" can fuse with the central tendon. Our central tendon appears to have three divisions or leaflets with the right leaflet being the largest and the left the smallest. Structurally, the central tendon comprises several planes of collagen fibers that intersect at various angles adding strength to the tendon.

The mechanical effects of force generated by muscle fibers in the different regions of our diaphragm muscle obviously depend on the specific origins and insertions of these fibers, as well as the varying external loads imposed by rib cage and abdominal displacement. In particular, there are marked differences between the costal and crural regions of the diaphragm muscle that lead some to suggest that these are actually two different muscles, with different embryonic origins and neural innervation [20, 21]. However, a number of studies have shown this to be incorrect. In detailed stud-

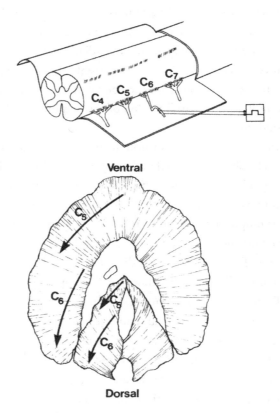

FIGURE 2: Segmental innervation of the cat diaphragm. Phrenic nerve axons, derived from the C4 spinal cord segment innervate more ventral aspects of the costal and crural regions. In comparison, phrenic axons from the C6 spinal cord segment innervate more dorsal aspects of both diaphragm regions. Used with permission from Sieck [89].

ies in rodents, Greer and colleagues clearly showed that all regions of the diaphragm muscle have similar embryonic origin [2, 4, 36]. Similarly, a systematic evaluation of the segmental innervation of different diaphragm muscle regions clearly demonstrated that the phrenic nerve, which emanates from the cervical spinal cord, innervates all diaphragm regions. The pattern of phrenic innervation in sternal, costal and crural regions is somatotopic with considerable overlap in the cervical segmental innervation of diaphragm fibers across regions. For example, phrenic nerve axons derived from higher cervical spinal cord segments innervate more ventral aspects of the costal and crural regions, whereas axons from lower cervical spinal cord segments innervate more dorsal aspects of both diaphragm regions [27, 89] (Figure 2).

2.2   INTERCOSTAL MUSCLES

In addition to our diaphragm muscle, which is responsible for lower rib cage expansion, we have other primary respiratory muscles that directly affect expansion (inspiration) and retraction (expiration) of our chest wall. Among these respiratory muscles, perhaps the most important are the intercostal muscles, which form three layers that occupy each of the intercostal spaces. Contraction of the intercostal muscles causes an elevation (expansion–inspiration) or retraction (expiration) of our ribs depending on the origin, insertion, and orientation of muscle fibers. This movement of our ribs is similar to raising and lowering a bucket handle, thereby expanding or constricting our thoracic cavity.

To understand the function of the intercostal muscles, it is first important to know the anatomy of our ribs. We have 12 ribs on each side that curve downward and anteriorly and comprise bony (lateral) and cartilaginous (medial) portions. Our 1st rib has a joint anteriorly with the manubrium, which is the superior part of our sternum. Our 2nd rib connects anteriorly at the joint between the manubrium and sternum bones—the manubriosternal synchondrosis. Our 3rd through 10th ribs connect anteriorly with the sternum with the cartilaginous portions of our 7th through 10th ribs fusing together before connecting to the sternum near the xyphoid process—near the xyphisternal junction. Our 11th and 12th ribs do not connect to the sternum at all and are sometimes called "floating" ribs. The shapes of the downward curving projections of our ribs appear as separate bucket handles, with varying curvatures, allowing different extents of upward (inspiration) and downward (expiration) motion.

The external intercostal muscles originate superiorly and laterally from the 1st through the 11th ribs and project obliquely downward and forward to insert on the lateral bony portions of our 2nd through 12th ribs. With contraction, the external intercostal muscles cause elevation of our ribs, thereby expanding the thoracic cavity leading to a decrease in intrathoracic and intrapleural pressures, causing inward airflow and lung inflation. The intercostal nerves innervate the external

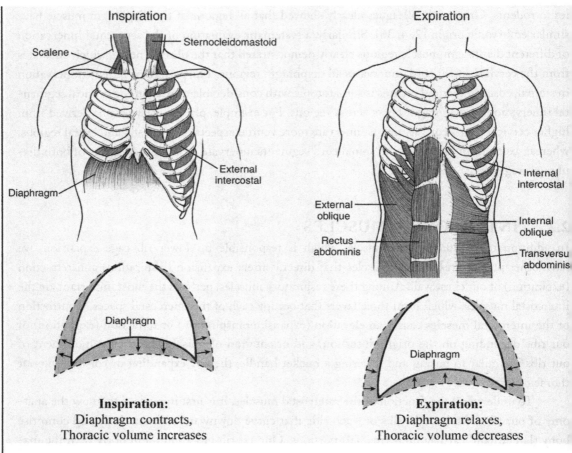

FIGURE 3: Inspiration during normal, quiet breathing is primarily generated by contraction of the diaphragm, external intercostal and parasternal intercostal muscles. During inspiration, the diaphragm contracts and moves caudally, causing an increase in the vertical dimension of the thoracic cavity and thus a decrease in intrathoracic pressure. Expiration is a passive process during quiet breathing, caused by passive elastic recoil of the lungs and chest wall. Active expiration requires recruitment of internal intercostal and abdominal muscles. Used with permission from Rhoades and Bell [86].

intercostal muscles segmentally with motor neurons located in the T1 through T11 spinal cord (Figure 3).

The internal intercostal muscles originate inferiorly from our 2nd through 12th ribs and project obliquely upward and medially to insert on our 1st through 11th ribs. Contraction of internal intercostal muscle fibers that originate laterally from the boney portion of the ribs below causes depression of the ribs above. This decreases the transverse dimensions of our thoracic cavity, leading to an increase in intrathoracic and pleural pressures and lung deflation. In contrast, contraction of internal intercostal muscle fibers that originate from the more medial cartilaginous portion of the ribs cause these ribs to lift upward, thereby expanding the thoracic cavity leading to an decrease in intrathoracic and pleural pressures and lung inflation. Thus, the mechanical action of these ventral internal intercostal muscles near the sternum is similar to that of the external intercostal muscles. For this reason, these ventral internal intercostal muscle fibers are often separated into a separate muscle group called the parasternal intercostal muscles. In fact, for the parasternal intercostal muscles, it may be helpful to consider that the origins and insertions of muscle fibers are actually reversed from that of the more lateral intercostal muscles. Contraction of these parasternal intercostal muscles elevates the ribs below, exerting a clear inspiratory function. Intercostal nerves from the T1 through T11 spinal cord segments also innervate the internal and parasternal intercostal muscles.

The deepest layer of the internal intercostal muscles are separated from the internal intercostal muscles by the neurovascular bundle, and thus these fibers form a distinct third muscle layer called the innermost intercostal muscles. This group of muscles is also sometimes called the transversus thoracis muscle and the subcostalis muscle. The transversus thoracis muscle is located anteriorly and originates from the posterior surface of the body of the sternum and xiphoid process, as well as from the sternal ends of the cartilaginous portions of our 7th through 10th ribs. These muscle fibers then diverge upward and laterally to insert into the lower borders and inner surfaces of the costal cartilages of the 2nd through 6th ribs. These muscle fibers form a thin layer of muscular and tendinous fibers that are continuous with fibers of the transversus abdominis muscle below. Contraction of transversus thoracis muscle fibers exert relatively minor mechanical effects on our chest wall, but may aid when our expiratory efforts become more forceful, together with the transversus abdominis muscle (see below).

The subcostalis muscles are typically found only in the lower part of our rib cage, with muscle fibers originating from the inner surface of a rib and then inserting on ribs located one to two segments lower. The orientation of subcostalis muscle fibers is in the same direction as internal intercostal muscle fibers. Contraction of these muscle fibers likely causes only a modest mechanical effect on our rib cage but as part of the innermost intercostal muscle group contraction of subcostalis muscle fibers can act as a synergist, together with the transversus thoracis muscle and lateral internal intercostal muscles, to assist when our expiratory efforts need to be more forceful. The internal intercostal nerves from T1 through T11 also segmentally innervate the innermost intercostal muscles.

The lateral portions of our intercostal spaces contain all three layers of intercostal muscles. In contrast, the ventral portions of our intercostal spaces contain only internal (parasternal) intercostal muscles [19]. As detailed above, contraction of intercostal muscles displaces the ribs upward or downward to affect inspiration or expiration, respectively. However, perhaps equally important, activation of intercostal muscles helps to stiffen and stabilize our rib cage, and thereby enhance diaphragmatic contraction and generation of transdiaphragmatic pressure.

In addition to our intercostal muscles, there are other muscles that insert on our rib cage and therefore may be involved in our respiratory efforts. For example, the scalene muscles originate from the transverse processes of the lower five cervical vertebrae and insert on the upper surfaces of our first two ribs [19]. Contraction of scalene muscle fibers causes elevation of our upper ribs and thus, these muscles contribute to inspiration, even during minimal inspiratory efforts [17, 84].

In the relatively rare case of bilateral diaphragm muscle paralysis, our remaining primary inspiratory pump muscles are capable of assuming adequate inspiratory pump activity to maintain resting ventilation [56]. However, it is far more common clinically that disease or some other condition compromises the ability of our diaphragm muscle to sustain adequate ventilation. For example, with chronic obstructive pulmonary disease (COPD), our diaphragm muscle becomes progressively weaker, and the contributions of the parasternal intercostal and scalene muscles become much more obvious and significant [16]. It should be noted that in addition to their role as respiratory pump muscles, the intercostal and scalene muscles also contribute to posture, head position, and stabilization of the shoulder and trunk during movements of the upper extremity [18, 47, 61].

2.3 ABDOMINAL MUSCLES

Generally, expiration is a passive process that results from passive elastic recoil of our chest wall and lungs following inspiratory-related inflation. However, under certain conditions, our expiration may become active involving generation of increased abdominal pressure by activation of abdominal muscles or increased intrathoracic pressure through activation of the internal and innermost intercostal muscles (see above).

Our abdominal muscles serve a variety of functions including movement and postural support of our trunk (together with the back muscles) as well as protection organs in our abdominal cavity. As with chest wall muscles, our abdominal muscles are organized into overlapping layers. The transversus abdominus muscle is the deepest abdominal muscle of the anterior and lateral abdominal wall. It lies in the same plane as the transversus thoracis muscle and is generally continuous with this muscle. It is a flat thin muscle and as is indicated by its name, muscle fibers run transversely across the anterior abdominal wall. The transversus abdominis originates broadly from the lateral portion of the inguinal ligament, from the superior portion of the iliac crest (top of the hip bone), and from the inner surfaces of the lower six ribs, where it interdigitates with diaphragm muscle fibers. Transversus abdominis muscle fibers insert anteriorly in a broad aponeurosis, together with muscle fibers of the internal oblique muscle forming the midline of our abdominal wall.

The internal oblique muscles form the next layer of abdominal muscles with muscle fibers also originating broadly from the thoracolumbar fascia of our lower back, the anterior portion of the iliac crest and the lateral half of the inguinal ligament. Internal oblique muscle fibers project upward toward the midline and insert at the inferior borders of our lower ribs (10th through 12th) and along the linea alba (abdominal midline). The orientation of internal oblique muscle fibers is perpendicular to those of the external oblique muscle.

Our external oblique muscle is the largest and most superficial of the three overlapping layers of muscles of the anterior and lateral abdominal wall. The external oblique muscle lies immediately anterior (more superficial) to the internal oblique muscle with muscle fibers running perpendicular to the internal oblique. External oblique muscle fibers also originate broadly from the inferior borders of our 5th through 12th ribs. The origin of external oblique muscle fibers interdigitates with the origin of serratus anterior (5th through 9th ribs) and the latissimus dorsi (10th through 12th ribs) muscle fibers. External oblique muscle fibers from the lowest ribs project vertically downward and insert into the iliac crest. External oblique muscle fibers originating from the upper ribs are also directed downward but also forward to insert into the aponeurosis that eventually merges with the linea alba. This aponeurosis of the external oblique muscle also forms the inguinal ligament, and a portion of the muscle contributes to the inguinal canal.

The rectus abdominus muscle is the most anterior of our abdominal muscles, but not the most superficial since the tendinous aponeurotic sheath extending from the external oblique muscles covers the rectus abdominus. Rectus abdominus muscle fibers project vertically downward on each side of the midline of anterior abdominal wall. The two parallel rectus abdominus muscles are separated by the linea alba, which is a band of connective tissue extending from the xiphoid process in the thorax to the pubic crest. The rectus abdominus muscle appears as separated into three portions divided by tendinous intersections of vertically oriented muscle fibers. The main function of the rectus abdominus together with other abdominal muscles is to bend our trunk forward (flexion); hence the exercise benefit of sit-ups. Importantly, the abdominal muscles also function, together with the back muscles, to stabilize the trunk and maintain posture.

As respiratory pump muscles, the abdominal muscles function only during forced expiration, together with the lateral portion of the internal and innermost intercostal muscles. However, activation of the abdominal muscles is also important in stiffening the abdominal wall, which facilitates the generation of transdiaphragmatic pressure by contraction of the diaphragm muscle (or external intercostal muscles). Thus, abdominal muscle "tone" is important in effective ventilation. This is demonstrated by the use of extrinsic abdominal binders to improve inspiratory muscle activity in certain pathological conditions such as spinal cord injury.

Activation of abdominal muscles is also an essential part of non-ventilatory behaviors that are important for expulsive clearance of our airways, e.g., when we cough or sneeze. During these non-ventilatory behaviors we generate near maximum transdiaphragmatic pressures. This involves maximum activation of the diaphragm muscle together with intercostal and abdominal muscles.

The simultaneous generation of maximum intrathoracic and intra-abdominal pressures, i.e., maximum transdiaphragmatic pressure, allows a powerful expulsion of foreign objects or noxious material from our airways.

With obesity, an accumulation of adipose tissue in the abdominal cavity can restrict the inspiratory movement of the diaphragm and chest muscles. In addition, the chest wall becomes less compliant, thereby increasing the external load for contraction of the diaphragm and inspiratory intercostal (external and parasternal intercostal) muscles making them less efficient. As a result, the work of breathing increases the diaphragm load and intercostal muscles may become more susceptible to fatigue. This may explain, at least in part, a condition called obesity hypoventilation syndrome (or Pickwickian syndrome), in which ventilation becomes inadequate, especially during sleep, where the condition may be exacerbated by obstructive sleep apnea. As a result, patients with severe obesity may have lower O_2 and higher CO_2 levels in their arterial blood.

2.4 ACCESSORY RESPIRATORY MUSCLES

The efficacy of our respiratory pump muscles is enhanced by stabilization of the chest wall via the mechanical effects contributed by the accessory respiratory muscles. There are a number of muscles that either originate from or insert on the chest wall. These muscles include the pectoralis major and minor, latissimus dorsii, sternocleidomastoid, triangularis sterni, serratus anterior, serratus posterior superior and inferior, upper trapezius, erector spinae (thoracic), iliocostalis lumborum, quadratus lumborum, and levator costarum muscles [19, 30].

These muscles act to stabilize and stiffen our chest wall during respiratory pump muscle activation, thereby enhancing the effect on intrathoracic pressure which in turn translates into altered intrapleural and intrapulmonary pressures and changes in air flow and lung volume. Our accessory respiratory muscles are relatively inactive during resting ventilation and are typically recruited only under conditions of increased inspiratory or expiratory efforts, e.g., during exercise or due to pathological conditions such as COPD [37, 75]. Thus, some may classify these accessory muscles in the same group as the major inspiratory and expiratory pump muscles.

· · · ·

CHAPTER 3

Airway Muscles

Airway muscles control the caliber of the conduits for air movement into and out of our lungs. There are four major parts of this conductive airway system: 1) the upper airways comprising the nasal and oral cavities and the nasal and oral pharynx; 2) the larynx and laryngeal folds or inlet; 3) the middle airways comprising the trachea and primary and more proximal bronchi; and 4) the lung airways comprising the more distal bronchi, bronchioles, and alveolar ducts. All of our airways must be patent in order to effectively move air into and out of our lungs. The airway muscles serve to actively regulate the diameter of our airways and thereby decrease resistance to airflow during both inspiration and expiration. Controlling the patency of our airways is also important in matching ventilation of alveoli to their perfusion and preventing aspiration of foreign objects or noxious material into our lungs. In particular, skeletal muscles in our upper airways and those controlling opening of the laryngeal inlet must exhibit activation patterns that are coordinated with those of our inspiratory and expiratory pump muscles to optimize air flow into and out of our lungs. A lack of coordination or insufficient activation of upper airway muscles (e.g., the genioglossus that causes tongue protrusion during inspiration) may lead to an obstruction of our airway. This is relatively common in a large proportion of the adult population and is diagnosed clinically as obstructive sleep apnea.

3.1 UPPER AIRWAY MUSCLES

Air enters the airways through either our nose (the preferred pathway) or mouth into the nasal and oral cavities. There are two nares or nostrils representing two channels separated by a septum for entry of air through the nose. The dilator naris muscle originates from the nasal notch of the maxillary bone and is inserted into the skin near the margin of the nostril. Contraction of the dilator naris muscle causes our nostrils to flare thereby opening the nares for air entry into our nasal cavity. The buccal branch of the facial nerve (cranial nerve VII) innervates the dilator naris muscle (Figure 4).

There are several muscles that affect opening or closing of our mouth by acting on the mandible or jaw. The lateral pterygoid muscle opens our mouth. It consists of two heads: 1) a superior head with muscle fibers originating from the infratemporal surface of the sphenoid bone and inserting into the articular disc and fibrous capsule of the temporal mandibular joint; and 2) an inferior

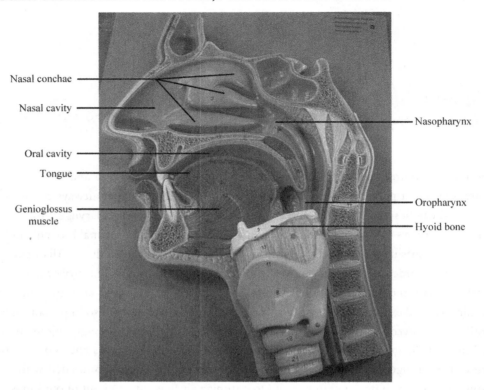

Nasal conchae

Nasal cavity

Oral cavity

Tongue

Genioglossus
muscle

Nasopharynx

Oropharynx

Hyoid bone

FIGURE 4: Sagittal view of the nasal and oral cavities. Modified with permission from http://www .flickr.com/photos/greenflames09/74296512/

head with muscle fibers originating from the lateral surface of the lateral pterygoid plate and inserting onto the neck of condoyle of the mandible. Contraction of the lateral pterygoid muscle depresses the mandible and thereby opens our jaw allowing air, food, and water to enter our mouth. It also functions when we chew food (mastication) together with three other muscles, the medial pterygoid, temporalis, and masseter muscles, all of which act to elevate the mandible and close the mouth—a much more forceful action. The mandibular nerve (cranial nerve V) innervates all four of these muscles of mastication.

Surrounding our mouth, the orbicularis oris muscle acts as a sphincter muscle, closing our mouth and puckering our lips when it contracts. The buccal branch of the facial nerve (cranial nerve VII) innervates the orbicularis oris muscle.

After air enters our nose or mouth, it then enters into the nasopharynx or oropharynx before passing through the laryngeal inlet into the larynx and finally into our trachea on its way to the lungs. Along the way, the external air is warmed and humidified. In our nasal cavity, the nasal conchae are structures that markedly increase the surface area so that the incoming air comes into contact with the highly vascularized mucous membrane that lines the entire nasal cavity. Exposure to this environment rapidly warms (or cools) and humidifies the incoming air. Our oral cavity also represents a large surface area of highly vascularized mucous tissue, where the incoming air is rapidly heated and humidified. Air coming through either the nose or mouth converges on the naso- or oropharynx that is also lined with highly vascularized mucous membrane for thermo-equilibration and humidification of the incoming air.

The patency of our airway is maintained during breathing by tightly coordinated co-activation of respiratory pump muscles and dilator (abductor—opening) or constrictor (adductor—closing) muscles of the upper airways. The main dilator muscle of our pharyngeal airway is the genioglossus muscle, which when activated causes our tongue to protrude forward away from the pharyngeal wall. This motion is very important to prevent the tongue from relapsing onto the pharyngeal wall and thus causing collapse of our airway. Insufficient activity of the genioglossus muscle during sleep occurs in may people and causes snoring and in some cases obstructive sleep apnea, where airflow into our lungs is completely blocked.

Our tongue (glossal) muscles are grouped as either intrinsic—entirely within the tongue, or extrinsic with fibers originating outside the tongue but inserting into the tongue. The intrinsic muscles of our tongue alter its shape during talking or swallowing, while the extrinsic muscles of our tongue reposition it toward or away from the pharynx; so together, they are much more important during breathing. There are four extrinsic muscles of the tongue that serve to protrude, retract, depress, and/or elevate our tongue. These muscles are the genioglossus (protrude), hyoglossus (depress), styloglossus (retract), and palatoglossus (elevate tongue but depress palate) muscles (Figure 5).

The genioglossus muscle is a fan-shaped muscle that is the largest extrinsic muscle of our tongue. Muscle fibers in the genioglossus muscle originate from the mental spine of the mandible, and then fan out dorsally to have a broad insertion from the hyoid bone in front to the entire dorsal part of our tongue behind. The genioglossus muscle is innervated by the hypoglossal nerve (cranial nerve XII). When fibers in our genioglossus muscle contract, they cause depression and protrusion of our tongue away from the pharyngeal wall. Thus, it is important for our genioglossus muscle to be activated during inspiration, and the coordination of genioglossus muscle activity with activation of the inspiratory pump muscles (e.g., diaphragm muscle, external intercostal muscles) is essential to avoid obstruction of our pharyngeal airway. Indeed, as mentioned above, sufficient activation of the genioglossus muscle during sleep may lead to obstructive sleep apnea. However, even with full activation, contraction of our genioglossus muscle alone may not be not sufficient to prevent narrowing of the upper airway in some cases, and obstructive sleep apnea may still result from other causes as well [62, 111].

Our hyoglossus muscle originates from the hyoid bone and its muscle fibers project upward to insert into the tongue. Contraction of our hyoglossus muscles causes the tongue to move downward

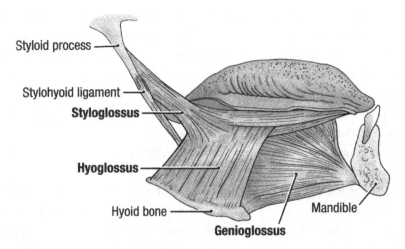

FIGURE 5: Lateral view of the tongue muscles, specifically the genioglossus, hyoglossus, styloglossus, and palatoglossus muscles. Used with permission from Agur and Dalley [1].

(depression) while also retracting the tongue. While the hyoglossus muscle probably has little respiratory function, it is important for us in phonation—speaking and singing. The hyoglossus muscle is also innervated by the hypoglossal nerve (cranial nerve XII).

The styloglossus muscle originates from the styloid process, a pointed protrusion of our temporal bone, and its fibers project downward and forward to insert into the dorsolateral aspects of the tongue, blending with fibers of the hyoglossus muscle. The styloglossus muscle also has little respiratory function, but aids in retracting and shaping the tongue when we swallow. The styloglossus muscle is also innervated by the hypoglossal nerve (cranial nerve XII).

The palatoglossus muscle originates from the anterior surface of our soft palate, and its muscle fibers project obliquely downward and forward to insert into the sides of the tongue. Contraction of palatoglossus muscle fibers causes elevation of the posterior portion of our tongue and depression of our soft palate. This mechanical action has little direct respiratory function, but by closing the oropharyngeal isthmus (see below), it functions when we swallow, an activity that must be coordinated with respiratory function to avoid aspiration. Unlike other extrinsic muscles of our tongue, the palatoglossus muscle is innervated by branches of the vagus nerve (cranial nerve X), similar to the pharyngeal constrictor muscles, which also functions during swallowing (see below).

The position of the hyoid bone is important in order to understand how upper airway muscles are involved in maintaining patency of our oropharynx. The hyoid bone does not directly articulate with any other bone; thus, it is highly mobile. During inspiration, it is important that our hyoid bone moves forward, otherwise resistance to airflow will increase. Several muscles insert at the hyoid bone, and contraction of these muscles causes important changes in the position of the hyoid bone. Three anterior muscles in the neck insert at the inferior border of our hyoid bone and are thus grouped as the infrahyoid (below the hyoid) muscles. The infrahyoid muscles are also called strap muscles of the neck because of their long flat shape. The origin of these infrahyoid muscles provides their specific names: 1) the thyrohyoid muscle has fibers that originate from the thyroid cartilage of the larynx. The thyrohyoid muscle appears as a continuation of another strap muscle, the sterno-thyroid muscle that originates from our sternum and inserts inferiorly at the thyroid cartilage. The thyrohyoid muscle is innervated by motor neurons in the C1 spinal cord, while the sternothyroid muscle is innervated segmentally by the C1–C3 spinal cord (by a plexus of nerves called the ansa cervicalis—see below). 2) The sternohyoid muscle originates from the superior portion of our sternum near the manubrium, as well as from the medial portion of the clavicle, and the sternoclavicular ligament. The sternohyoid muscle is also segmentally innervated by C1–C3 (ansa cervicalis). 3) The omohyoid muscle consists of inferior and superior portions or bellies that are separated by a tendon attached to our clavicle. Muscle fibers in the inferior belly of the omohyoid muscle originate from our shoulder; hence, its name, which derives from the Greek term for shoulder—omos. Specifically, these muscle fibers originate from the upper border of the scapula. Muscle fibers in the inferior belly of the omohyoid muscle project forward and slightly upward in our lower neck to insert into the middle tendon. Muscle fibers in the superior belly of our omohyoid muscle originate from this middle tendon and change direction to project almost vertically upward in our neck to insert into the hyoid bone. The omohyoid muscle is also innervated segmentally by motor neurons in the C1–C3 spinal cord.

Contraction of the infrahyoid muscles causes depression and posterior displacement of our hyoid bone and depression of the larynx. The mechanical effects of elevating (suprahyoid muscles—see below) and then depressing the hyoid bone must be coordinated during swallowing to avoid aspiration of food or water into our lungs (Figure 6).

There is another group of muscles that originate from or have central tendons that attach to the hyoid bone and then project upward. The suprahyoid muscles include the digastric, stylohyoid, geniohyoid, and mylohyoid muscles. Generally, most of these suprahyoid muscles do not have a direct respiratory function. However, simultaneous contraction of the sternohyoid and geniohyoid muscles moves our hyoid bone in the anterior direction causing dilation of the upper airway, a mechanical effect that is important during inspiration. In contrast, contraction of the other suprahyoid muscles causes our hyoid bone to move upward or elevate; thereby causing the oropharynx to widen when we swallow (see below). More importantly, when the suprahyoid muscles elevate the hyoid bone, the larynx is pulled upward folding the epiglottis and thus closing the glottis or laryngeal inlet

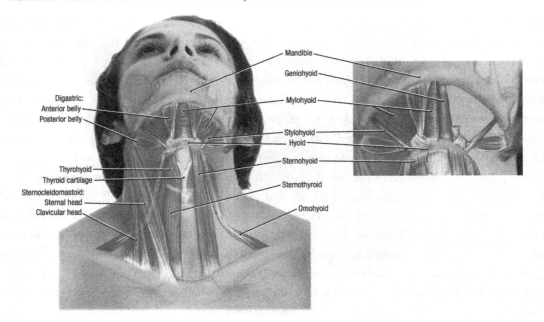

FIGURE 6: Anterior view of the hyoid muscles, specifically the infrahyoid muscles: omohyoid, sternohyoid, sternothyroid, and thyrohyoid muscles. Also shown are the suprahyoid muscles: digastric, stylohyoid, geniohyoid, and mylohyoid muscles. Used with permission from Agur and Dalley [1].

(see below). Although the suprahyoid and infrahyoid muscles function during different phases of the motor pattern of swallowing, their activities must be coordinated with inactivity of our inspiratory pump muscles to avoid aspiration of food and water into our lungs when we swallow.

The pharynx extends from the base of our skull down to the separation of the larynx (see below) and esophagus. It is located behind (posterior to) our nasal and oral cavities and in its lower portion behind our larynx. For most of its length, the pharynx forms a common pathway for swallowing food and water and for airflow during respiration. The pharynx is often separated into three parts: the nasopharynx, oropharynx, and laryngopharynx (Figure 7).

The nasopharynx is a continuation of our nasal cavity, with the nasal choanae representing the junction between our nasal cavity and the nasopharynx. In the back of our throat, the nasopharynx merges with the oropharynx at a narrow passage or strip of tissue (isthmus) that is bounded anteriorly by the soft palate, laterally by the palatopharyngeal arches, and posteriorly by the wall of the pharynx. The tonsils are lymphoid tissue embedded in the mucous membrane of the posterior wall of the naso- and oropharynx (see below), and when enlarged, they may cause respiratory obstruction. In the lateral walls of the nasopharynx are openings of our auditory tubes, where another

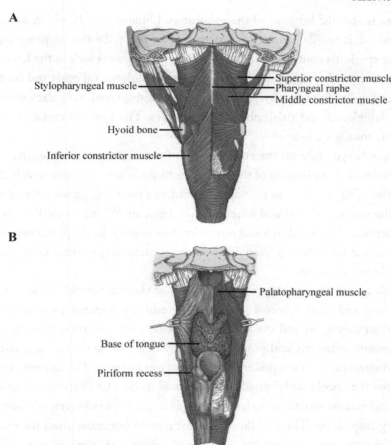

A

Stylopharyngeal muscle——

Superior constrictor muscle
Pharyngeal raphe
Middle constrictor muscle

Hyoid bone——

Inferior constrictor muscle——

B

Palatopharyngeal muscle

Base of tongue——

Piriform recess——

FIGURE 7: A) Structure of the muscular wall of the pharynx. Three constrictor muscles overlap and join at the midline. B) Posterior view of an opened pharynx showing the relationship of the palatopharyngeus muscle, base of the tongue and piriform recess of the laryngopharynx. Used with permission from Fritsch and Kuehnel [29].

set of lymphoid tissue called tubal tonsils is located. The auditory tubes connect the nasopharynx to the tympanic cavity of our ears and serve to equalize the pressure of the external air with that in the tympanic cavity of our inner ears. When we swallow, the pharyngeal isthmus is closed by muscular action to avoid reflux into the nasal cavity, but the auditory tube is opened to equalize pressure.

The oropharynx extends from the soft palate to the top of the epiglottis—part of the larynx (see below). The oropharynx communicates with the oral cavity and is separated by an isthmus bounded superiorly by the soft palate, laterally by the palatoglossal arches, and inferiorly by the tongue. This area is also the location of the palatine and lingual tonsils, which together with nasopharyngeal and tubal tonsils form a ring of lymphatic tissue. In the oropharynx, the mucous membrane of the epiglottis continues onto the base of the tongue as well as the lateral walls of the pharynx. This space on either side of midline is called the epiglottic vallecula and contains taste receptors. The lateral walls of the oropharynx comprise the palatoglossal and palatopharyngeal arches, overlying the palatoglossal and palatopharyngeal muscles. The space between the two arches is where the palatine tonsils are located.

The laryngopharynx extends the entire length of our larynx, which is located in front, from the top of the epiglottis to the bottom of the cricoid cartilage, where it connects with the esophagus. The glottis or inlet of the larynx is at the superior border of the laryngopharynx and represents the point at which the passage of food and water diverges from air. When we swallow food and water, the laryngeal inlet must be closed to avoid aspiration (see below). In the space between the laryngopharynx behind and the larynx (arytenoid and cricoid cartilages) in front is the piriform recess, where food may become lodged.

In the walls of our pharynx, there are two layers of skeletal muscles: 1) an external, circular layer of constrictors and 2) an internal layer of longitudinally oriented muscles that elevate the pharynx—the stylopharyngeus and the palatopharyngeus. There are three pharyngeal constrictor muscles that originate anteriorly and project posteriorly in a circular overlapping fashion to insert into a broad tendonous raphe in the posterior midline of our pharynx. The inferior constrictor muscle originates from the cricoid and thyroid cartilages, and its circular fibers are continuous with the esophagus. The inferior constrictor muscle functions as a sphincter to prevent air from entering the esophagus during inspiration. The middle constrictor muscle originates from the hyoid bone, and the superior constrictor muscle originates from the mandible and sphenoid bone. Contraction of the pharyngeal constrictor muscles facilitates swallowing of food and water. The pharyngeal branch of the vagus nerve (cranial nerve X) innervates each of the pharyngeal constrictor muscles.

The palatopharyngeus muscle originates from the boney palate and inserts into the thyroid cartilage and the side of the pharynx. The pharyngeal branch of the vagus nerve (cranial nerve X) innervates the palatopharyngeus muscle similar to the pharyngeal constrictor muscles. The stylopharyngeus muscle originates from the styloid process of the temporal bone and then passes between the superior and middle constrictor muscles to insert with fibers of the palatopharyngeus muscle. The stylopharyngeus muscle is innervated by the glossopharyngeal nerve (cranial nerve IX).

The major function of our pharyngeal muscles is swallowing, which is a complex motor behavior requiring coordination of tongue, suprahyoid, infrahyoid, and pharyngeal muscle contraction as food or water passes from the mouth through the oropharynx and then laryngopharynx to the esophagus on its way to the stomach. When we swallow, the nasopharynx and laryngeal inlet must be closed to avoid reflux into the nose and airways, respectively. Thus, neural control of the motor function of swallowing must be coordinated with the neuromotor control of respiration.

The negative pressures we generate during inspiration will cause our pharyngeal walls to collapse unless there is opposing tone generated by pharyngeal muscle contraction (stiffening). In addition, to counter the collapse of the pharyngeal wall during inspiration, contraction of our airway dilator muscles—the genioglossus, geniohyoid, and sternohyoid muscles—is essential. Thus, activity in these dilator muscles is synchronized with that of the inspiratory pump muscles. Respiratory-related activation of our inferior pharyngeal constrictor muscle is also quite important, but for a different reason. Contraction of the inferior pharyngeal constrictor muscle exerts a sphincter function blocking airflow into the esophagus.

3.2   LARYNGEAL MUSCLES

Our larynx is a complex structure that separates our pharynx and upper airway above and our trachea below. It is involved in breathing, vocalization, and protecting the lower airway and lung against food aspiration. The larynx contains the vocal folds, which represent the borders of the laryngeal inlet for movement of air into and out of our lungs. Controlling the vocal folds is essential for phonation; when we speak or sing, we must adjust the tension on our vocal folds in order to manipulate the pitch and volume of sound we make. The vocal folds are located just above the point where the lower portion of the pharynx splits to form the esophagus. For airflow, the larynx connects the lower portion of the pharynx with the trachea.

The skeleton of the larynx is cartilaginous, consisting of three single cartilage structures: the epiglottis, thyroid and cricoid cartilages; and three paired cartilage structures: the arytenoid, corniculate, and cuneiform cartilages. Although not part of the larynx, the hyoid bone is connected to it. The interior of the larynx is separate into the supraglottis, glottis, and subglottis (Figure 8).

At the top of our larynx is the epiglottis, which consists of elastic cartilage covered with a mucous membrane. The epiglottis projects obliquely upwards behind the tongue and the hyoid bone. Importantly, our epiglottis guards the entrance to the larynx, called the glottis, with the opening between the vocal folds or laryngeal inlet. Normally, our epiglottis points upward during breathing, but when we swallow, the hyoid bone is elevated by contraction of the suprahyoid muscles, causing the larynx to move upward. This causes the epiglottis to fold downward covering our glottis and thereby preventing food and water from going into the trachea but directing it down the laryngopharynx and into the esophagus instead.

The prominent thyroid cartilage forms the anterior and lateral portions of the larynx. In the midline, there is prominent notch (or incisure—prominent "Adam's apple" in men), and at the lateral borders, there are superior and inferior cornus (horns). The inferior cornu articulates with the cricoid cartilage below, the lower border of which represents the most superior portion of the trachea. The thyrohyoid membrane connects the thyroid cartilage with the hyoid bone. The thyrohyoid membrane comprises a distinct median thyrohyoid ligament and a lateral thyrohyoid ligament. Inferiorly, there is also a median cricothyroid ligament.

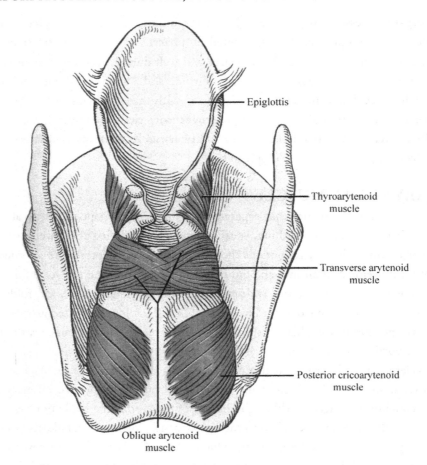

Epiglottis

Thyroarytenoid
muscle

Transverse arytenoid
muscle

Posterior cricoarytenoid
muscle

Oblique arytenoid
muscle

FIGURE 8: Posterior view of the larynx. The prominent epiglottis projects upward and during swallowing folds down to close the glottis or laryngeal inlet for air. During inspiration, the posterior cricoarytenoid muscles abduct or open the laryngeal inlet to allow airflow. These are the only laryngeal abductor muscles and thus serve an important role in controlling airway patency during inspiration. Used with permission from Fritsch and Kuehnel [29].

The paired arytenoid cartilages are pyramidal shaped and represent the interior portion of the larynx, where the vocal folds attach. The arytenoids articulate with the lateral part of the cricoid cartilage forming cricoarytenoid joints at which they can rotate and tilt forward or backward. Changes in position of our arytenoid cartilages result in movement of our vocal folds, thus changing the aperture of the laryngeal inlet and the tension on our vocal folds. At the tip of each of the arytenoid cartilages is a small but distinct conical-shaped cartilage, the corniculate cartilage. The paired corniculate cartilages do not appear to have any distinct function apart from the arytenoids. Also situated at the top of the arytenoid cartilage on each side is the small cuneiform cartilage, which forms part of the aryepiglotic fold, forming the lateral portion of the epiglottis and helping to support the vocal folds.

The muscles of our larynx are separated into extrinsic and intrinsic groups. These laryngeal muscles act to stabilize the thyroid cartilage and control the opening (abduction) and closing (adduction) of our laryngeal inlet (vocal folds); thus modulating airflow to and from the trachea below. The control of airflow through the laryngeal inlet is important not only for our breathing but also for vocalization. In the larynx, the only muscle that opens the laryngeal inlet is the posterior cryoarytenoid muscle. Contraction of the posterior cricoarytenoid muscle causes rotation of the arytenoid cartilages and abduction or separation of the vocal cords, thereby facilitating airflow. Other non-dilator and adductor muscles of the larynx play important roles in non-respiratory actions of the upper airways, e.g., swallowing and phonation.

3.3   AIRWAY SMOOTH MUSCLES

The trachea begins at the lower end of the cricoid cartilage of the larynx. In humans, the trachea is about 2.5 cm in diameter and extends for a length of 10–16 cm before dividing into two primary bronchi. Along its length, there are ~18 incomplete C-shaped cartilaginous rings that strengthen the anterior and lateral sides of the trachea and thereby protect the airway. The dorsal or back wall of the trachea is membranous. Smooth muscle, the trachealis muscle, connects the cartilaginous rings and when contracted, reduces the diameter of the inner lumen of the trachea causing an increase in airway resistance. The incomplete cartilaginous rings also allow our trachea to partially collapse when food passes down the esophagus, which is located immediately posterior to the trachea (Figure 9).

The trachea divides into two primary (or main) bronchi at the level of the 4th and 5th thoracic vertebrae just behind the sternum. The right main bronchus subdivides into three branches, each going to a separate lobe of the right lung. In contrast, the left main bronchus divides into two branches going to the two lobes of the left lung. These secondary, lobar bronchi divide into tertiary bronchi, each of which supplies a bronchopulmonary segment, forming a distinct anatomical lung division separated by a connective tissue septum. Hyaline cartilaginous rings are present in the larger bronchi, but gradually disappear in the smaller bronchi as branching progresses.

The tertiary, segmental bronchi further divide into many primary bronchioles (4th branch point), which then divide again into terminal bronchioles (5th branch point) before giving rise to multiple respiratory bronchioles (6th branch point). Bronchioles subsequently divide into a number of alveolar ducts, with 5–6 alveolar sacs associated with each alveolar duct. Each alveolus forms a basic functional unit where gas exchange occurs within our lungs.

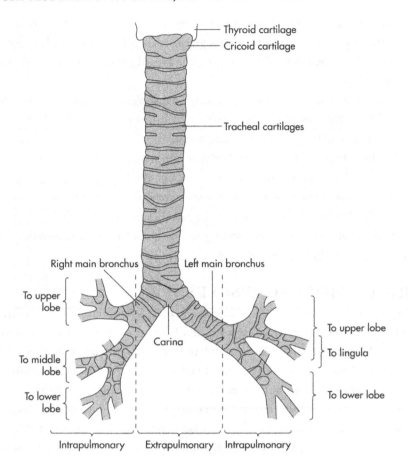

FIGURE 9: Structure of the trachea and major bronchi. The trachea is about 2.5 cm in diameter and extends for a length of 10–16 cm before dividing into two primary bronchi before entering the lungs. The primary bronchi then divide into a number of lobular bronchi within the lung. Modified with permission from Beachey [6].

Throughout our lungs, the innermost portions of bronchial airways are lined with mucous membrane and epithelial cells. Along the way, the mucous membrane undergoes a transition from a ciliated pseudo-stratified columnar epithelium to a simple cuboidal epithelium and finally to a simple squamous epithelium that is present in the alveolar ducts and alveoli. This simple squamous epithelial layer presents only a minimal diffusion barrier for gas exchange with the associated pulmonary blood vessels.

As mentioned above, our trachea, bronchi, bronchioles, and alveolar ducts are also lined with airway smooth muscle. Unlike skeletal muscle, the ultrastructure of smooth muscle is not organized into sarcomeres. On one hand, this complicates the analysis of contractile mechanisms (see below); however, it also provides greater adaptability, especially with respect to the force–length relationship of smooth muscle (see below). In addition, the neural innervation of our smooth muscle is completely different from that found in skeletal muscle with no well-defined neuromuscular junctions (synapses—see below). Instead, a plexus of nerves from the autonomic nervous system innervates smooth muscle cells indirectly. As a result, our airway smooth muscle cells are exposed to a variety of neural agonists and antagonists influencing contractility. The balance between agonist and antagonist effectors establishes the "tone" of airway smooth muscle contraction.

Smooth muscle is found lining the walls of all the major (and minor) conduits within our body-blood vessels (e.g., vascular smooth muscle), the urinary tract (e.g., uterine smooth muscle), the gastrointestinal tract (e.g., gastrointestinal smooth muscle), and the airways (e.g., airway smooth muscle). The structure and function of smooth muscle in these organ systems is similar in many respects, but there are important differences in neural regulation and mechanical responses that precludes extrapolation of smooth muscle properties from one organ system to another.

In contrast to skeletal muscle and cardiac muscle, the structure and function of our smooth muscle is substantially different. There are major differences in cytoskeletal structure, contractile proteins, regulation of contraction, and excitation–contraction coupling. These differences will be highlighted in the sections below (Figure 10).

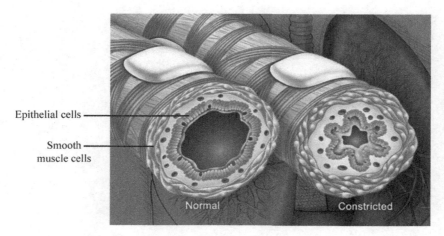

Epithelial cells

Smooth muscle cells

Normal   Constricted

FIGURE 10: Depiction of the layers of airway muscle, namely the outer smooth muscle layer and the inner epithelial cells. In conditions such as asthma, the airways are chronically inflamed and constricted. Modified with permission from http://www.health68.com

Generally, smooth muscle cells have a fusiform shape (20 to 500 μm in length), with a single nucleus, which is markedly different from skeletal muscle fibers that are multinucleated. The cytoskeletal structure of smooth muscle cells is also greatly different compared to the highly organized crystalline structure of skeletal and cardiac muscle (see below). As a result, smooth muscle is far more adaptive and can maintain active contractile function across a much wider range of lengths (length adaptation) [5].

·   ·   ·   ·

CHAPTER 4

Muscle Structure and Function

It is essential to understand the basic physiology of skeletal and smooth muscle to understand differences in the function of various respiratory muscles. Skeletal muscle is composed of varying numbers of multinucleated cells called myofibrils or muscle fibers. In contrast, smooth muscle cells have a single nucleus. The essential contractile proteins—actin and myosin—are similar between skeletal and smooth muscle, but the regulation of actin–myosin interactions and the mechanisms mediating excitation–contraction coupling are markedly different. The range in the mechanical performance of our skeletal muscle is generally much greater than that found in our smooth muscle. However, the energetic requirements of skeletal muscle contraction are also generally much greater than that for smooth muscle contraction. These differences in functional properties between our skeletal and smooth muscle will be discussed in the following sections.

4.1   SARCOMERE STRUCTURE OF SKELETAL MUSCLE

The contractile proteins within our skeletal muscle fibers are well organized as sarcomeres that provide the striated appearance of our skeletal and cardiac muscles. In contrast, contractile proteins within our smooth muscle cells are not organized into well-defined sarcomeres, which complicates physiological assessment of mechanical performance of these muscles. In both skeletal and smooth muscle, thick filaments contain the myosin molecule that interacts (binds) with the actin molecule of the thin filament to form cross-bridges that are the essential units of force generation and contraction—the two primary functions of muscle fibers. The composition of these contractile proteins varies across different muscle types, and these differences in contractile protein composition impart different mechanical and energetic properties (e.g., different skeletal muscle fiber types—see below). The skeletal muscle fiber type composition of our respiratory muscles allows a wide range of ventilatory and non-ventilatory behaviors. In contrast, the range of functional motor behaviors in our airway smooth muscles is much more limited.

The basic structural unit of a skeletal muscle fiber is the sarcomere, comprising thick (myosin) and thin (actin) filaments aligned in an interdigitating, crystalline structure. The sarcomere itself is bounded at each end by a dense Z-disc (Z-line) from which the actin filaments project toward the midline, while thick filaments are situated in the middle of the sarcomere. The Z-disc runs perpendicular to the filaments and connects neighboring sarcomeres, creating a functional unit that permits transmission of lateral and longitudinal force during contraction. The dimension of each sarcomere is approximately 1 µm in diameter and 2.5 µm in length (Z-line to Z-line). The overlap between thick and thin filaments determines the number of cross-bridges that can be formed during muscle contraction. The thick filament has a relatively fixed length of ~1.6 µm, while the thin filament length ranges between 1.0 and 1.3 µm and is species and fiber type-dependent. During muscle fiber contraction, the intrinsic lengths of both the thick and thin filaments do not change, but the binding of the myosin head to actin pulls the Z-line of the sarcomere toward the midline, thus increasing the overlap between thick and thin filaments. The number of sarcomeres in series can vary but generally does not exceed ~20 mm (~8000 sarcomeres in series) (Figure 11).

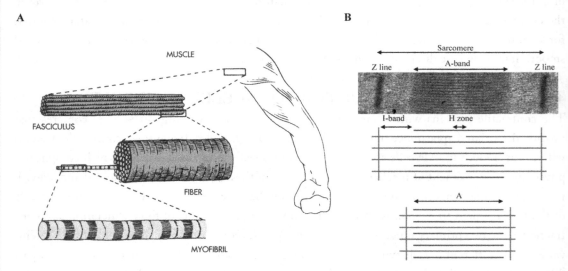

FIGURE 11: A) Depiction of the sarcomere, the essential structural unit of skeletal muscle. The sarcomere comprises thick (myosin) and thin (actin) filaments, which slide past each other during contraction. Thick filaments are comprised of myosin, titin, and C-protein, and constitute the A-band. Thin filaments are composed of actin, tropomyosin, and troponin, and are anchored in the Z-line. The basic unit of skeletal muscle contraction is a cross-bridge formed by the attachment of the myosin head to the actin filament. Used with permission from Berne et al. [7]. B) The ultrastructure of a skeletal muscle sarcomere can be seen with transmission electron micrography. The length of the A-band is constant during changes in muscle and sarcomere length, while length of the I-band decreases.

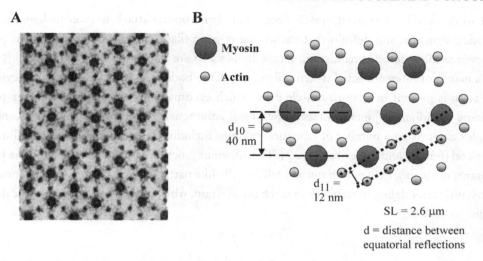

A

B

Myosin

Actin

$d_{10} =$
40 nm

$d_{11} =$
12 nm

SL = 2.6 μm

d = distance between
equatorial reflections

FIGURE 12: A) Transmission electron micrograph of a cross-section of a skeletal muscle fiber showing the thick and thin filament lattice structure. B) Schematic depiction of a cross-section of a muscle fiber, showing a lattice of thick (myosin) and thin (actin) filaments. Surrounding each thick filament there are 6 thin filaments with fixed intervals between thick and thin filaments.

The structural organization of the sarcomere appears crystalline with a fixed stoichiometry between the number of thick and thin filaments. This crystalline structure of the sarcomere is evident in electron microscopic images and by X-ray diffraction. In cross-sections of skeletal muscle fibers, each myosin filament is surrounded by six actin filaments, which are further surrounded by six myosin filaments. The spacing between each myosin and actin filament is relatively fixed in a lattice structure. Thus, this arrangement creates a double hexagonal array forming a myofilament lattice. Understanding the thick and thin filament spacing in this myofilament lattice is key to understanding the interactions between thick and thin filaments during skeletal muscle force generation and contraction. The filament lattice provides stability to the sarcomere and balances radial and axial forces placed upon it. Since the sarcomere is encompassed by the sarcolemma, muscle fiber osmolarity shifts caused by changes in ion concentrations across the membrane can result in osmotic compression of the lattice and change the spacing between thin and thick filaments. Under hypertonic conditions, lattice spacing and muscle force generation are decreased. Conversely, under hypotonic conditions, lattice spacing increases and may lead to a decrease in force (Figure 12).

4.2   INFRASTRUCTURE OF SMOOTH MUSCLE CELLS

As mentioned above, our smooth muscle cells are not structurally organized into sarcomeres. Instead, in smooth muscle cells, α-actin filaments attach to dense bodies that are rich in α-actinin similar to the Z-disc in skeletal muscle fibers. The dense bodies attach to intermediate filaments comprising vimentin and desmin that anchor the α-actin filaments to each other and the plasma membrane to allow force transmission. Dense bodies also are associated with cytoskeletal β-actin. Thus, a matrix of interconnected α-actin filaments, dense bodies, intermediate filaments, cytoskeletal β-actin is present in smooth muscle cells, which eventually becomes attached to the plasma membrane via adherens junctions (also called focal adhesions). The adherens junctions consist of a complex comprising a number of structural proteins including α-actinin, vinculin, paxillin, talin, cytoskeletal β-actin, and integrins [38, 39]. The adherens junctions are scattered around the plasma membrane, encircling the smooth muscle cell in a rib-like pattern. These focal adhesion complexes are constantly remodeling in response to mechanical strain, which is mediated in part by focal adhesion kinases (FAK).

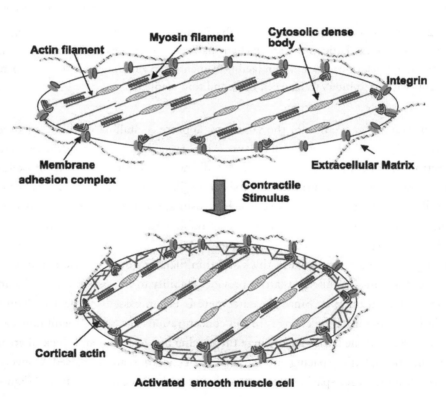

FIGURE 13: Depiction of the structure of smooth muscle cells. A matrix of interconnected actin filaments, dense bodies, and intermediate filaments is present in smooth muscle cells, which attach the contractile proteins to the cell membrane and extracellular matrix via adherens junctions and integrin proteins. Used with permission from Gunst and Zhang [40].

Airway smooth muscle cells also contain regions of membrane invagination called caveolae. In many respects, these caveolae are similar to transverse tubules in skeletal muscle fibers (see below). Currently, there is intense interest in the functional role of caveolae within airway smooth muscle cells.

When actin and myosin interact to form cross-bridges (see below), force is generated and transferred to the sarcolemma through the dense bodies, intermediate filaments, adherens junction, and integrins that attach the contractile proteins to the plasma membrane. In turn, the plasma membrane of airway smooth muscle cells attaches to the extracellular matrix of the lung via specific interactions of the integrin proteins, thereby providing external loading of smooth muscle contraction. A very unique feature in airway smooth muscle cells is that during contraction, there is a reorganization of the cytoskeleton and contractile proteins to optimize force generation. Furthermore, there can be a change in the amount of contractile proteins present in airway smooth muscle cells. For example, it appears that myosin and actin filaments can polymerize during smooth muscle activation, thereby changing contractile protein content and the number of cross-bridges that can form (see below). Accordingly, the stochiometric relationship between actin and myosin filaments can change even during the course of a single contraction (Figure 13).

4.3 MYOSIN CROSS-BRIDGE AND FORCE GENERATION

For both skeletal and smooth muscle fibers, the basic functional unit underlying force generation and contraction is the cross-bridge, representing the cross-linking of a single myosin head with the actin filament. Within a sarcomere, six actin filaments surround each myosin filament, and the position of myosin heads spiral around the thick filament to align with the position of the actin filaments. There are ~300 myosin heads per myosin filament. Within a typical human diaphragm muscle fiber that has a cross-sectional area of ~3,000 μm^2 there would be ~2 million myosin filaments and ~600 million myosin heads. Assuming a specific force of ~30 N per cm^2 for diaphragm muscle fibers, this would equate to ~0.5 to 1.0 pN per myosin head (cross-bridge).

The force generated by a skeletal muscle fiber depends on the number of cross-bridges formed in parallel. With an increase in skeletal muscle fiber cross-sectional area, there is an increase in the both the number of sarcomeres in parallel and thus, the number of myosin heads in parallel that can form cross-bridges and contribute to force generation. For this reason, muscle force is often normalized by cross-sectional area (specific force). Relating this to our daily lives, we know that as our muscles increase in overall size (increase in the cross-sectional areas of muscle fibers—hypertrophy), they are able to generate more force. Thus, weight lifters work on increasing the size of their muscles through weight training.

During force generation and contraction, cross-bridges pull actin filaments toward the midline of the sarcomere. Thus, at the Z-discs of the sarcomere, the force vectors generated by

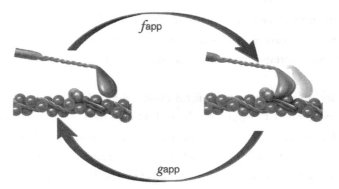

FIGURE 14: Cross-bridge cycling between two functional states—strongly bound and unbound—results in force generation and contraction. f_{app}—apparent rate constant for cross-bridge attachment; g_{app}—apparent rate constant for cross-bridge detachment. Used with permission from Sieck and Regnier [105].

cross-bridge formation are in opposite directions toward the midline of the sarcomere. Accordingly, within a skeletal muscle fiber, force generation is dependent on the number of strongly bound force generating cross-bridges in parallel per half-sarcomere (Figure 14).

The actual number of cross-bridges that can form depends on the myosin head (myosin heavy chain) content per half-sarcomere (n, maximal number of potential cross-bridges that can form) and the proportion of these myosin heads that actually form strongly bound force-generating cross-bridges (α_{fs}). This proportion of myosin heads that actually bind to actin and form cross-bridges is influenced by the overlap of thick and thin filaments (underlying the force–length relationship of skeletal muscle fibers) and the myoplasmic Ca^{2+} concentration (underlying the force–Ca^{2+} relationship of muscle). Finally, each cross-bridge contributes a unit of force, and the total force generated by a skeletal muscle fiber will thus depend on the average force produced per cross-bridge (f). Thus, the force (F) generated by a muscle fiber can be determined by the following equation:

$$F = n \cdot \alpha_{fs} \cdot f$$

The fraction of strongly bound cross-bridges (α_{fs}) during muscle fiber activation can be estimated by measuring muscle fiber stiffness [33, 57]. In these measurements, muscle fibers are maximally activated and then high frequency (~2 kHz) sinusoidal length oscillations of 0.01% of optimal muscle length (Lo) are imposed. This small amplitude length change is used to avoid disruption of cross-bridge binding. The high frequency oscillation far exceeds the cross-bridge cycling rate; thus the recoil force induced by the length perturbations reflects the number of strongly bound cross-

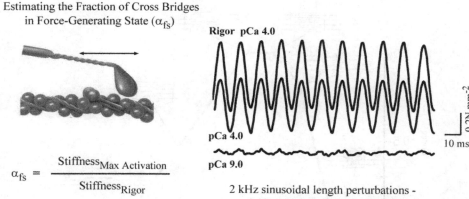

Estimating the Fraction of Cross Bridges
in Force-Generating State (α_{fs})

Rigor pCa 4.0

pCa 4.0

pCa 9.0

0.2N mm^{-2}

10 ms

$$\alpha_{fs} = \frac{\text{Stiffness}_{\text{Max Activation}}}{\text{Stiffness}_{\text{Rigor}}}$$

2 kHz sinusoidal length perturbations -
0.1% optimal sarcomere length

FIGURE 15: The fraction of strongly bound cross-bridges (α_{fs}) during muscle fiber activation can be estimated by comparing muscle fiber stiffness during Ca^{2+} activation to that during a rigor condition (Ca^{2+} activation in the absence of ATP—assumed to be maximum cross-bridge formation and stiffness). Stiffness of a single diaphragm muscle fiber is measured by imposing small amplitude length perturbations (~0.1% of optimal length at 2 kHz) during activation at varying Ca^{2+} concentrations (pCa 9.0—relaxation to pCa 4.0—maximum activation). Used with permission from Han et al. [42].

bridges. To approximate the maximum number of strongly bound cross-bridges, muscle fibers are activated (by exposure to high Ca^{2+} solutions) in the absence of ATP; a rigor condition where cross-bridges remain strongly bound. During maximum Ca^{2+} activation, approximately 80% of recruitable cross-bridges are strongly bound, and this fraction does not differ across different fiber types. The a_{fs} measure by this method varies with myoplasmic Ca^{2+} concentration from a pCa^{2+} (inverse log of Ca^{2+} concentration) of 9.0 (complete relaxation) to a pCa^{2+} of 4.0 reflecting maximum Ca^{2+} activation [40]. The dependency of muscle fiber stiffness on myoplasmic Ca^{2+} concentration precisely mirrors the dependence of muscle force generation (Figure 15).

The force generated by a skeletal muscle fiber also depends on sarcomere length and the extent of overlap between thick and thin filaments. There is an optimal sarcomere length for muscle fiber force generation (Lo). If sarcomere length is either shorter or longer than Lo, force generation decreases. This relationship, known as the force–length relationship of skeletal muscle, reflects the fraction of strongly bound cross-bridges. If skeletal muscle fiber stiffness is measured during maximum Ca^{2+} activation, maximum stiffness occurs at Lo (Figure 16).

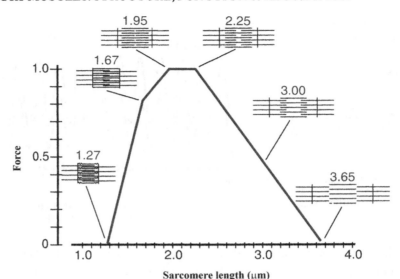

FIGURE 16: The force generated by a muscle fiber depends on the extent of overlap between the thick and thin filaments, such that maximal force is achieved at optimal sarcomere length. The force–length relationship defines the fraction of strongly bound cross-bridges at varying sarcomere lengths. Used with permission from Rhoades et al. [86].

4.4   MUSCLE FIBER SHORTENING VELOCITY

In addition to force generation, the other major function of our muscle fibers is to shorten and thereby exert force on an external load. The force (or load)–velocity relationship of muscle is another important relationship in characterizing the mechanical properties of muscle fibers. The velocity at which a muscle shortens depends inversely on the external load imposed. Thus, the force–velocity relationship is anchored at zero load where maximum unloaded shortening velocity occurs, and at a maximum load where the intrinsic force generated by the muscle fiber is offset by the external load and no shortening occurs. As cross-bridges cycle between attached and unattached states (see below), cross-bridges remain attached for a finite amount of time before detaching. The rate of cross-bridge cycling, and thus the time cross-bridges remain attached, is dependent on external load. Thus, with increasing velocity of shortening, the number of attached cross-bridges decreases. The inverse is also true that when velocity decreases, more cross-bridges are attached and generate force. Thus, the muscle fiber stiffness–velocity relationship mirrors the force–velocity relationship (Figure 17).

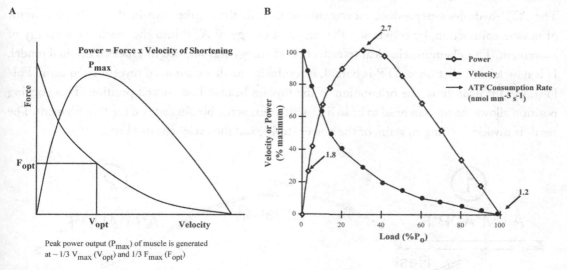

A)

Power = Force x Velocity of Shortening

P_{max}

Force

F_{opt}

V_{opt} | Velocity

Peak power output (P_{max}) of muscle is generated
at ~ 1/3 V_{max} (V_{opt}) and 1/3 F_{max} (F_{opt})

B) Power / Velocity / ATP Consumption Rate (nmol mm^{-3} s^{-1})

Velocity or Power (% maximum)

Load (%P_0)

FIGURE 17: A) The force–velocity–power relationship of muscle. The velocity at which a muscle shortens is inversely related to the imposed external load or force against which it must contract. The power produced by a muscle fiber is the product of the force it generates and its velocity of shortening. B) ATP consumption rate varies with power output in single muscle fibers. The heat generated by a muscle (as a result of ATP consumption) increases with work or power performed—first demonstrated by W. O. Fenn. Maximum ATP consumption in a muscle fiber corresponds to peak power (or work) output (termed the Fenn effect). Modified with permission from Sieck and Prakash [102].

4.5   CROSS-BRIDGE CYCLING

Cross-bridges cycle between states of attachment and detachment, and this cross-bridge cycling requires energy in the form of ATP hydrolysis. Thus, during the cross-bridge cycle, there is a complex enzymatic reaction that transduces the energy stored within ATP into force and mechanical work. The importance of the cyclical interactions between myosin and actin was first recognized in the sliding filament theory of muscle proposed by Andrew Huxley [48], in which the cross-bridge cycle was described as a two-state model. In this simple model, cross-bridges exist in one of two states: 1) myosin heads strongly bound to actin (attached cross-bridges) and generating force, or 2) myosin heads not bound to actin (cross-bridges detached) with no force generation. In this cross-bridge cycling model, muscle force and contraction occurs during the power-stroke phase, as the myosin head binds to actin, bending at the junction between the head and neck regions of myosin molecule. The amount of force and shortening depends on the external load opposing the intrinsic force generated. Subsequent detachment of the cross-bridge is dependent on the hydrolysis of ATP.

The ATP-hydrolysis-dependent movements of myosin thus make myosin the molecular motor of muscle contraction, by converting the chemical energy of ATP into the mechanical energy of movement. This chemomechanical transduction of energy was implied in Huxley's original model. It is now known that once ATP is bound, it results in the dissociation of myosin from actin. Following ATP hydrolysis, the orientation of the myosin head is in a cocked position. This cocking position allows the myosin head to be in line with a next actin-binding site on the thin filament. The result is myosin binding to actin, or the power stroke, and the cycle repeats (Figure 18).

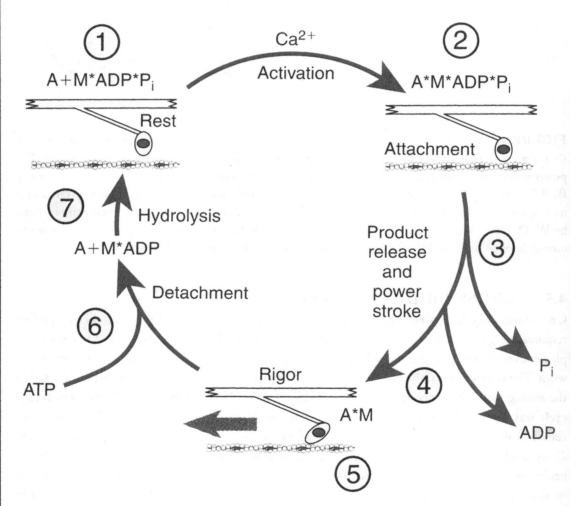

FIGURE 18: Steps in the myosin cross-bridge cycle. Cross-bridge detachment and cycling depends on ATP binding to myosin head and ATP hydrolysis. Shortening velocity is dependent on cross-bridge cycling rate. Used with permission from Rhoades and Bell [86].

As mentioned above, the velocity of shortening of muscle fibers is directly related to cross-bridge cycling rate, and the external load imposed on a muscle. The fact that force generation requires cross-bridge attachment, which in turn is inversely related to cross-bridge cycling rate, underlies the hyperbolic relationship between force and velocity of shortening. Assuming a simplified two-state model of cross-bridge cycling where cross-bridges cycle between a strongly bound force-generating state to an unbound state, there would be two apparent rate constants to describe cross-bridge cycling: f_{app}—a forward rate constant for cross-bridge attachment, and g_{app}—a reverse rate constant for cross-bridge detachment, which is dependent on ATP hydrolysis [8, 9]. In this model, the fraction of cross-bridges in a strongly bound force-generating state (α_{fs}) would be characterized as:

$$\alpha_{fs} = f_{app} / (f_{app} + g_{app})$$

4.6 MUSCLE FIBER ENERGETICS

If in the model presented above it is assumed that one molecule of ATP is consumed per cross-bridge cycle, then ATP consumption in a skeletal muscle fiber with a total length of b half sarcomeres in series can be determined by:

$$\text{ATP consumption} = n \cdot b \cdot \alpha_{fs} \cdot g_{app}$$

Wallace O. Fenn observed that the heat produced by a muscle during activation increases with increasing work (or power) output of the muscle (termed the Fenn effect) [25]. The power produced by a muscle fiber is the product of the force it generates and its velocity of shortening. Work is the power produced by our muscle over a period of time. We now know that as power or work of a muscle fiber increases, energy requirements (ATP consumption) increase. Peak power or work of a muscle fiber occurs at approximately 33% of maximum velocity and maximum force [102]. It has been shown that maximum ATP consumption in a muscle fiber corresponds to peak power output [98] (Figure 19).

In our smooth muscle cells, the same basic relationships hold true. ATP consumption drives cross-bridge cycling between attached and detached states. However, with isometric activation of airway smooth muscle cells, the rate of ATP hydrolysis changes with time. Initially, the rate of ATP hydrolysis is higher and then it markedly slows as activation is sustained, even though length and external load do not change. This dynamic slowing of cross-bridge cycling kinetics and the rate of ATP hydrolysis does not occur in our skeletal muscle fibers under isometric conditions. Therefore, this dynamic change in energy consumption is a unique property of our smooth muscle cells and lead to the formulation of the "latch bridge" hypothesis. The latch-bridge hypothesis suggests that

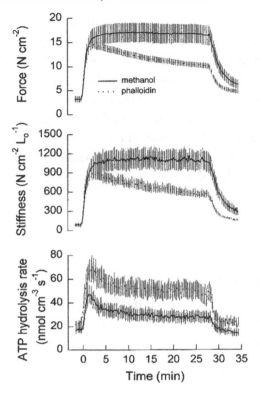

FIGURE 19: Effect of f-actin stabilization with phalloidin on the time course of isometric force generation, stiffness and ATP hydrolysis rate in canine tracheal smooth muscle strips. Phalloidin reduces the ability of smooth muscle to maintain isometric force and stiffness, but results in an increased rate of ATP hydrolysis. In airway smooth muscle, dynamic actin polymerization and remodeling are necessary for attaching the contractile proteins to the cell membrane. With phalloidin this cytoskeletal remodeling is blocked so cycling cross-bridges are not attached to the cell membrane. As a result, the internal load imposed on contractile proteins is reduced, which affects an increase cross-bridge cycling rate and ATP consumption. Used with permission from Jones et al. [54].

the duty cycle for cross-bridge attachment (and thus cross-bridge cycling rate) in smooth muscle cells is regulated by the extent of myosin light chain phosphorylation (see below). However, it is also possible that the change in cross-bridge cycling rate and ATP hydrolysis rate in smooth muscle cells reflects a time-dependent increase in internal loading of actin–myosin filaments with cytoskeletal remodeling. Accordingly, as smooth muscle cells are initially activated, intracellular $[Ca^{2+}]$ increases and cross-bridges are formed (see below for mechanism of excitation–contraction coupling). However, the dense bodies and intermediate filaments within smooth muscle cells are not yet attached to the plasma membrane, and thus the contractile apparatus is relatively unloaded resulting in a faster cross-bridge cycling rate and ATP consumption rate. As the cytoskeleton of the smooth muscle cell remodels and dense bodies become anchored to the plasma membrane, internal loading of the contractile proteins increases, thereby slowing cross-bridge cycling rate and decreasing ATP hydrolysis rate [28, 54].

4.7   LOADING OF MUSCLE FIBERS

For a muscle fiber, a load is force that opposes the contraction of muscle fibers. In its simplest form, the forces generated by cross-bridges in each half sarcomere of a muscle fiber pull the Z-disc toward the midline of the sarcomere. Thus, at each end of the sarcomere, there are external loads opposing the force generated by cross-bridges. Typically, muscle fibers have a site of origin that is fixed and a point of insertion where an external load is moved. This simplifies our analysis of external load and muscle fiber contraction. The external loads placed on skeletal muscle fibers can be broadly categorized as either: 1) dynamic, elastic loading, or 2) static, resistive loading. With dynamic, elastic loading the force exerted by the external load changes with time. For example, if our muscle fiber pulls on a rubber band the external load increases as the rubber band is stretched by muscle fiber contraction. In contrast, with static resistive loading the force exerted by the external load does not change. Static resistive loading occurs commonly during resistance training or weight lifting.

For inspiratory pump muscles, the external loads imposed by the lung and chest wall represent dynamic, elastic loading due to their elastic recoil evident during passive expiration. There are many natural and pathological conditions where the elastic properties of the lungs and chest wall can change [65]. For example, during early postnatal development, the elastance of the lungs and chest wall is very low (they are highly compliant, which is the reciprocal of elastance). Thus, to inflate the lungs during inspiration, the diaphragm and other inspiratory pump muscles in neonates must generate less force compared to the adult where the recoil forces of the lung and chest wall are much greater (increased elastic loading). As we age, the elastance of the lung and chest wall progressively increases (i.e., they become stiffer); thus, the inspiratory pump muscles must generate greater relative forces to sustain ventilation (see below). The abdominal wall and cavity can also present an external elastic load for the inspiratory pump muscles. For example, during pregnancy, abdominal wall displacement and increased pressure in the abdominal cavity will oppose the downward displacement of the diaphragm muscle during inspiratory contractions. Thus, to sustain ventilation, the diaphragm muscle must generate increased force (transdiaphragmatic pressure).

Static resistive loading of the inspiratory pump muscles is fairly rare. In conditions of airway occlusion such as obstructive sleep apnea, the diaphragm muscle would be exposed to an increased resistive load.

We have all experienced friends in the weight room of our gym who use weight lifting to increase external loading as a means to induce skeletal muscle fiber hypertrophy and thus an increase in total force-generating capacity. Remember, the force generated by a muscle fiber depends on the number of cross-bridges formed in parallel per half sarcomere. With an increase in fiber cross-sectional area, there is an increase in myosin heads and thus the number of cross-bridges that can form and contribute to force. Conversely, we have all seen examples of the result of skeletal muscle unloading when the arms or legs of our friends are placed in a cast for a period of time, and as a result, their muscle fibers atrophy and become weaker [115].

For respiratory pump muscles, changes in external loading result from changes in the elastance (reciprocal of compliance) of the lung and chest wall. For example, in patients with chronic obstructive pulmonary disease (COPD), the compliance of their lungs increases. Thus, the external dynamic elastic loading of their inspiratory muscles is actually reduced. However, because of the lower elastic recoil of their lungs, patients with COPD have a problem deflating their lungs during expiration. Thus, the residual air volume in their lungs after expiration is increased and the diaphragm muscle remains flattened at a shorter length. Both the shorter length and reduced curvature of the diaphragm diminishes its mechanical performance, and as a result additional force must be generated to inflate the lungs. Thus, the impact of disease related changes in external loading of the diaphragm muscle is sometimes quite complex. With the altered loading of the diaphragm muscle associated with COPD, there is also a decrease in the myosin heavy chain content per half sarcomere and the force generated per cross-bridge [72]. Thus, although there is not a decrease in diaphragm muscle fiber cross-sectional area, there is a marked reduction in specific force that makes the diaphragm much weaker in these COPD patients [63].

Internal loading of longitudinal force transmission by sarcomeres can also occur due to contractile structures that are situated in parallel or in series with sarcomeres. For example, increased collagen or other extracellular matrix elements that are positioned in parallel with myofibrils can increase the passive stiffness of our muscle and thereby affect force generation. Passive elastic properties of muscle can resist lengthening or shortening of muscle fibers. As we age, there is a change in the quality of the collagen in the extracellular matrix surround muscle fibers with increased cross-linking between collagen fibers. This age-related change increases the passive stiffness of our muscles and affects their mechanical performance [35]. When our limbs are immobilized in casts for long periods of time, the extracellular of our muscles becomes stiffer. The same is true when our muscles are injured during exercise. Contractures are common in neuromuscular disorders and may also increase passive muscle stiffness. Such changes in passive mechanical properties of muscle are evident as a decrease in joint range of motion due to limitations from within the muscle tissue. The limitation due to increased passive stiffness of the extracellular matrix can decrease the force generating capacity and decrease shortening velocity.

Changes in external and internal loading of smooth muscle certainly occur, even during the course of normal activation. Airway smooth muscle cells are tethered to the extracellular matrix of the lung, which is a compliant tissue. The recoil forces of lung parenchyma will depend on the extent of lung inflation; thus, external load on the airway smooth muscle will vary during the course of the respiratory cycle. As mentioned above, both the contractile proteins and the cytoskeleton within airway smooth muscle cells remodel during activation and in response to mechanical strain [39]. This will lead to changes in the internal loading of the contractile proteins as well as changes in the transmission of force to the plasma membrane and ultimately the extracellular matrix. These are very complex interactions that are as yet unresolved (Figure 20).

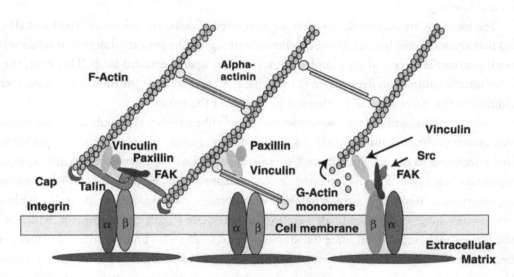

FIGURE 20: Structure of integrin and cytoskeletal adherens junctions in airway smooth muscle. Adherens junctions consist of structural proteins including: α-actinin, vinculin, paxillin, talin, cytoskeletal β-actin, and integrins. An increase in internal loading of airway smooth muscle results from the attachment of contractile proteins to the cell membrane via adherens junctions. The adherens junctions also contain integrin receptor proteins that tether the contractile proteins to the extracellular matrix of the lung, which imposes an elastic external load. Used with permission from Gunst et al. [39].

4.8   MUSCLE FIBER FORCE TRANSMISSION

Within skeletal muscle fibers, the force generated by the power stroke of the cross-bridge is transmitted toward the midline of the sarcomere pulling the Z-disc at one end. Thus, force is transmitted longitudinally from one sarcomere in series to another. From the sarcomeres, force is transmitted longitudinally through the relatively non-compliant tendons attached to the origins and insertions of muscle fibers, which are themselves attached to bones across a joint. Skeletal muscle torque is the representation of the force required to move or rotate a corresponding axis at a joint. For example, walking requires muscle force development via joint motion at the ankle, knee, and hip. As originally recognized by Charles Sherrington, the contractions of opposing muscle groups across a joint (e.g., flexors and extensors) are coordinated in a reciprocal pattern. Accordingly, as the flexors at the elbow joint (e.g., biceps brachii and brachioradialis muscles) are active the extensors (e.g., triceps brachii muscle) are inactive. This pattern of reciprocal activation of agonist and antagonist muscles is typical of rhythmic motor patterns such as walking, running, and breathing. Thus, during breathing, when inspiratory muscles are active, expiratory muscles are inactive. However, co-activation of opposing muscles does occur and can provide stabilization at a joint and the control of external forces and loading.

Tendons that attach muscles to bone are composed of collagen (primarily type I and III collagen) that are non-compliant and have great tensile strength. The primary function of tendons is to transmit contractile force and thus implement movement against external loads. Therefore, the efficacy of muscle contraction depends on the tendinous attachments (at the muscle and bone levels), in addition to the molecular and mechanical properties of the tendon.

Force is also transmitted transversely within muscle fibers to the cell membrane or sarcolemma. Costameres are subsarcolemmal structures within skeletal muscle fibers that connect peripherally located sarcomeres to the cell membrane. Costameres comprise a complex of dystrophin-associated glycoproteins (e.g., dystroglycans and sarcoglycans) that are aligned with the Z-disc. Within muscle fibers, costameres couple the force generated by sarcomeric cross-bridges to the cell membrane; thereby contributing to the coordinated control of sarcomere length and alignment, which is essential for optimal longitudinal force transmission. Costameres also link the internal cytoskeleton of the muscle fiber to the extracellular matrix via coupling to extracellular proteins such as collagen and laminin. Such coupling to the extracellular matrix may be very important in strain-dependent signaling within skeletal muscle fibers. In many respects, this transverse coupling of skeletal muscle fibers resembles the coupling of contractile proteins in smooth muscle to the extracellular matrix via adherens junctions (see above). Costamere dysfunction underlies certain types of myopathies (e.g., Duchenne muscular dystrophy).

In airway smooth muscle cells, force transmission is via an interaction with the extracellular matrix of the lung (see above). To mediate this interaction, smooth muscle cells have specific elastin and collagen receptors (integrin receptors) to mediate interactions between the adherens junctions and integrin proteins with these extracellular matrix proteins. While these integrin receptors mediate force transmission between the contractile proteins of the airway smooth muscle to the extracellular matrix, the tension developed in the extracellular matrix of the lung during lung inflation is transmitted back to the airway smooth muscle cell. Thus, the elastic load imposed by the lung extracellular matrix during lung inflation can initiate remodeling of the cytoskeleton as well as the contractile proteins (e.g., actin and myosin polymerization). This remodeling can be both acute and long-term and affect the contractile function of airway smooth muscle cells.

·   ·   ·   ·

CHAPTER 5

Muscle Fiber Proteins

5.1 MUSCLE CONTRACTILE PROTEINS

The two major contractile proteins are actin and myosin that interact to form cross-bridges (see above). There are multiple classes of both actin and myosin that serve a variety of functions other than muscle force generation and contraction. For example, there are as many as 18 classes of myosin and at least 6 types of actin that provide a variety of muscle and non-muscle motor functions as well as cytoskeletal structural support within cells. Within muscle, the interaction of myosin II and α-actin is responsible for force generation and contraction.

Within each sarcomere of a skeletal muscle fiber, there is a highly organized pattern of thick and thin filaments representing the contractile apparatus of the skeletal muscle and giving muscle a striated appearance. The thick filament comprises myosin II, a hexameric contractile protein consisting of two myosin heavy chains (MyHC—~220 kDa) and four myosin light chains (2 MyLC20—~20 kDa and 2 MyLC17—~17 kDa). Each of the heavy chains consists of an N-terminal head domain and a C-terminal tail domain. The myosin light chains serve to bind the heavy chains in the "neck" region between the head and tail (Figure 21).

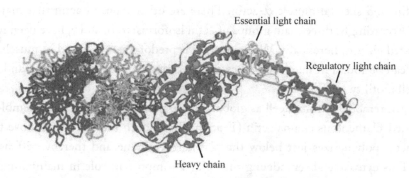

Essential light chain

Regulatory light chain

Heavy chain

FIGURE 21: Structure of skeletal myosin subfragment-1 showing the heavy chain and its two associated light chains domains. Used with permission from Hopkins et al. [46].

Myosin II in smooth muscle comprises a myosin heavy chain (SM-MyHC) encoded by a single gene (MyHC11). However, there are splice variants of this gene that result in four distinct isoforms of SM-MyHC. Also, smooth muscle may contain non-contractile classes of myosin that are not involved in force generation and contraction, but may play a role in cytoskeletal remodeling and trafficking of intracellular proteins and organelles (e.g., mitochondrial movement).

The two myosin heavy chains wrap around each other to form a double helix, creating the tail region of myosin [85]. The globular head region is on the end of the heavy chains, where each myosin head has two of the myosin light chains that control the head during contraction. The tail portion of myosin is positioned toward the midline of the sarcomere and the heads are directed inward toward the center of the thick filament. This allows for a part of each myosin molecule to create an angled arm that can attach to actin in the thin filament and form a cross-bridge. The two myosin heavy chains consist of both head N-terminal and tail C-terminal domains. The N-terminal domain includes regions for the binding site to actin and the enzymatic activity of ATP hydrolysis. The thick filament also contains titin and C-protein, which aid in filament stabilization.

The MyLC20 is also known as the regulatory light chain and its role in skeletal muscle contraction is currently debated, although there are phosphorylation sites that when phosphorylated appear to affect cross-bridge cycling kinetics and shortening velocity. In contrast, in smooth muscle, MyLC20 phosphorylation plays an active and direct role in regulating cross-bridge formation (see below under excitation–contraction coupling). In all types of muscle, the MyLC17 is referred to as an "essential" light chain, although its exact function is unclear. However, the MyLC17 likely contributes to the structural stability of the myosin head.

The thin filament is composed of three main proteins, α-actin, tropomyosin, and troponin. Alpha actin is a globular protein (~40 kDa) that contains the binding sites for the myosin heavy chain. Alpha-actin is also expressed as distinct genetic isoforms: smooth muscle α-actin, cardiac muscle α-actin, and skeletal muscle α-actin. There are other types of actin that may exist within muscle cells. Accordingly, three main groups of actin isoforms: α, β, and γ, have been identified and can be separated electrophoretically. While α-actin is predominantly found in muscle, the β and γ actin groups can be found in many cell types as constituents of the cytoskeleton and as mediators of internal cell motility.

Actin can remodel within a cell as globular actin monomers (G-actin) assemble into helical double stranded filamentous chains actin (F-actin). In smooth muscle, in response to contractile agonists, β-actin polymerizes just below the plasma membrane and thereby stiffens the smooth muscle cell. This cytoskeletal remodeling may play an important role in maintaining mechanical tension in smooth muscles in an energy efficient manner.

As mentioned above, α-actin and myosin are the major contractile proteins. The ratio of α-actin to myosin varies across different muscle types. In smooth muscle the ratio of α-actin to myosin ranges 2:1 to 10:1. In contrast, in skeletal muscle, the stoichiometric relationship between α-actin and myosin is relatively fixed at ~6:1, reflecting the lattice structure of skeletal muscle.

5.2 ACTIN REGULATORY PROTEINS IN SKELETAL MUSCLE

In addition to α-actin, the thin filament in skeletal muscle fibers contains tropomyosin and troponin that are important in regulating the availability of the binding site for the myosin heavy chain on actin. In skeletal muscle, a single tropomyosin molecule spans seven actin molecules and in the non-activated muscle, it overlays the myosin head binding sites on actin and is maintained in this position by troponin T (tropomyosin binding troponin) and troponin I (inhibitory troponin). When Ca^{2+} is released from the sarcoplasmic reticulum, it binds to troponin C (Ca^{2+} binding troponin), causing a conformational change in the troponin complex that displaces tropomyosin exposing the myosin binding sites on actin. The binding of a myosin head facilitates displacement of adjacent tropomyosins, allowing additional myosin heads to bind (cooperativity). Once myoplasmic Ca^{2+} is reduced, tropomyosin once again overlays the myosin binding sites on actin. Thus, together with the troponin complex, tropomyosin functions to implement Ca^{2+} regulation of cross-bridge attachment and force generation in a skeletal muscle fiber (Figure 22).

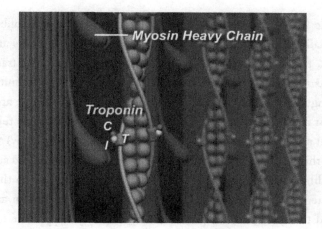

FIGURE 22: Animation showing the interaction of thick and thin filaments during cross-bridge cycling. The troponin complex functions to regulate cross-bridge attachment—the binding of the myosin head to the actin filament. This regulatory process begins with Ca^{2+} binding to troponin-C, which causes a conformational change in troponin-I (inhibitory) and troponin-T (bound to tropomyosin) and thus displaces tropomyosin from the myosin binding site on actin. http://tinyurl.com/Sieck-Cross-Bridge

Thin filaments in smooth muscle do not contain troponin; therefore, the regulation of cross-bridge formation is quite different from that in skeletal muscle. In smooth muscle calmodulin assumes the regulatory role linking an increase in myoplasmic Ca^{2+} with cross-bridge attachment to actin and force generation (see below, excitation–contraction coupling in smooth muscle). Tropomyosin is present in airway smooth muscle, spanning seven actin monomers, and as in skeletal muscle, removal of steric hindrance of the actin binding site is critical for cross-bridge attachment and force generation: although through a mechanism still unknown.

5.3   EXCITATION–CONTRACTION COUPLING IN MUSCLE

The mechanisms underlying excitation–contraction coupling vary considerably between skeletal muscle and airway smooth muscle. Yet, in both muscle types, it is essential to have an understanding of the mechanisms underlying excitation–contraction coupling, and this provides the essential control of muscle contraction by external influence (e.g., neural activation). The major difference in excitation–contraction coupling is that the regulatory site resides at the thin filament in skeletal muscle but at the regulatory myosin light chain (MyLC20) in smooth muscle. Another major difference between skeletal and smooth muscle is the regulation of myoplasmic Ca^{2+} in response to stimulation (excitation). In skeletal muscle fibers, the mechanisms linking activation (neural) and an increase in myoplasmic Ca^{2+} concentration are simpler and well understood. In contrast, in smooth muscle, there are multiple mechanisms linking agonist excitation and an increase in myoplasmic Ca^{2+} concentration.

Skeletal muscle fibers are larger and longer and comprise many myofibrils with sarcomeres arranged both in parallel and in series. Myofibrils in skeletal muscle fibers are surrounded by an intracellular network, the sarcoplasmic reticulum, which serves to store intracellular Ca^{2+} for release into the myoplasm surrounding the sarcomeric contractile proteins to initiate contraction (see above). Along the length of the muscle fiber membrane (sarcolemma), there are deep invaginations of the membrane that run transversely to the sarcolemma, called transverse tubules (or T-tubules). Action potentials that are initiated by neuromuscular transmission (see below) are propagated along the sarcolemma, and this depolarization is transmitted down the t-tubule and activates dihydropyridine receptors. The dihydropyridine receptors are Ca^{2+} channels located on the plasma membrane of the t-tubule situated just above the sarcoplasmic reticulum that contain another type of Ca^{2+} release channel called the ryanodine receptor channel. The dihydropyridine receptor channels interact with the sarcoplasmic reticulum to trigger release from the sarcoplasmic reticulum through ryanodine receptor channels. The two receptors are close in relation to each other but are not in direct contact in the absence of depolarization. Depolarization changes the conformation of the dihydropyridine receptor, and as this occurs, there is indirect contact of the two receptors, activating the ryanodine receptor. Thus, there is a dependence on the mechanical coupling of dihydropyridine and ryanodine receptors for muscle contraction. In skeletal muscle fibers, this process occurs in the absence of extracellular Ca^{2+}. The initial Ca^{2+} release from the sarcoplasmic reticulum causes additional Ca^{2+} release through a process called Ca^{2+}-induced Ca^{2+} release causing a rapid flooding of the myoplasmic space with Ca^{2+} (Figure 23).

Genetic mutations of the ryanodine receptor can affect sarcoplasmic reticulum Ca^{2+} release. For example, a mutation of the ryanodine receptor channel causes malignant hyperthermia, where in response to inhaled anesthetics, there is uncontrolled Ca^{2+} release from the sarcoplasmic reticulum, leading to muscle contracture, increased metabolic demands, hyperthermia, and possibly death.

As mentioned above, the amount of force generated by a muscle fiber is dependent upon myoplasmic Ca^{2+} concentration, which affects the fraction of strongly bound cross-bridges (α_{fs}— see above). The sensitivity of force to myoplasmic Ca^{2+} concentration varies across fiber types;

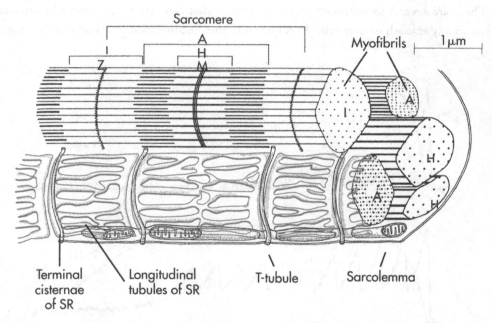

FIGURE 23: Depiction of myofibrils surrounded by the sarcoplasmic reticulum, which serves to store intracellular Ca^{2+}. Muscle fiber action potentials, initiated by neuromuscular transmission, are propagated along the sarcolemma. Depolarization is then transmitted passively down the T-tubule (transverse invaginations of the plasma membrane). Within the T-tubule, depolarization activates dihydropyridine receptors that trigger release of Ca^{2+} from the sarcoplasmic reticulum into the myoplasm. The released Ca^{2+} then binds to troponin-C triggering cross-bridge formation and force generation. Used with permission from Berne et al. [7].

fibers expressing $MyHC_{slow}$ have greater sensitivity to myoplasmic Ca^{2+} concentration than fibers expressing $MyHC_{2A}$, $MyHC_{2X}$, and $MyHC_{2B}$ isoforms. The sensitivity of force generation to myoplasmic Ca^{2+} concentration is commonly represented as the myoplasmic Ca^{2+} concentration at which 50% of maximal force is generated (pCa^{2+}_{50}) (Figure 24).

Smooth muscle fibers are much smaller and shorter; thus, the active propagation of electrical excitation is not nearly as critical, although it still can occur through gap junctions between smooth muscle myocytes. However, it is equally important to link excitation to the release of intracellular Ca^{2+} from internal sarcoplasmic reticulum stores. This can occur through electrical coupling via voltage-dependent Ca^{2+} channels in response to neurotransmitter-initiated depolarization the plasma membrane, or via triggered release of Ca^{2+} from intracellular stores (e.g., from the sarcoplasmic reticulum) [73]. Thus, compared to skeletal muscle fibers, the regulation of myoplasmic Ca^{2+} concentration in airway smooth muscle cells is far more complex and involves the interaction of multiple pathways.

There are several neurotransmitters (mediators) that can affect an increase in myoplasmic $[Ca^{2+}]$ in airway smooth muscle cells. In response to these neurotransmitters or agonists (e.g., ace-

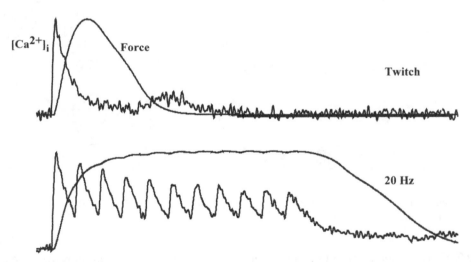

FIGURE 24: Ca^{2+} and force responses in a skeletal muscle fiber triggered by electrical stimulation. With a single stimulus, there is a transient increase in myoplasmic Ca^{2+} concentration that precedes and triggers the force response. The transient Ca^{2+} response is shorter in duration compared to the force response. With repetitive stimulation at 20 Hz, there is summation of both the myoplasmic Ca^{2+} and force responses. The extent of summation depends on the duration and relaxation of both Ca^{2+} and force responses.

tylcholine—ACh, histamine, serotonin—5HT, etc.), an increase in myoplasmic [Ca^{2+}] may occur via multiple mechanisms including Ca^{2+} influx through ligand-gated, G-protein-coupled or voltage-gated receptor channels [77, 82, 99]. Moreover, G-protein-coupled receptor channels can trigger an increase in production of a second messenger (e.g., ionositol triphosphate—IP_3 or cyclic ADP-ribose—cADPR) that in turn triggers release of Ca^{2+} from the sarcoplasmic reticulum through IP3-receptor or ryanodine receptor channels, respectively [3, 55, 78].

In response to an increase in myoplasmic [Ca^{2+}] in airway smooth muscle cells, cross-bridge attachment does not occur until the myosin heads have been activated by MyLC20 phosphorylation. The phosphorylation of MyLC20 is mediated by an enzyme called myosin light-chain kinase (MLCK) that is activated by Ca^{2+} binding to calmodulin. Specifically, activation of MyLS20 results from phosphorylation of a serine on position 19 (Ser19) on the MLC20 light chain. This phosphorylation results in a conformational change that increases the angle in the neck domain of the myosin heavy chain that facilitates binding of the myosin head to the actin filament. There are a number of cell signaling pathways that regulate MyLC20 dephosphorylation as well, e.g., the Rho kinase pathway (Figure 25).

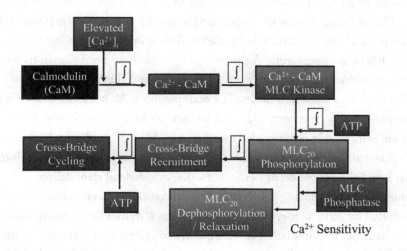

FIGURE 25: Excitation–contraction coupling in airway smooth muscle cells involves multiple steps. Agonist-induced activation of ligand-gated or voltage-gated receptors channels triggers an increase myoplasmic Ca^{2+} concentration. The increased Ca^{2+} binds to calmodulin (CaM) that activates myosin light chain kinase (MLCK) mediating phosphorylation of MyLC20, which in turn allows cross-bridge formation. Dephosphorylation of MyLC20 is regulated myosin light chain phosphatase (MLCP) and plays a role in modulating the Ca^{2+} sensitivity of force generation in airway smooth muscle.

Other cell signaling pathways may be involved in regulating β-actin polymerization. Thus, neurotransmitters may also play a role in cytoskeletal remodeling and force maintenance in airway smooth muscle cells. As mentioned above, it may also be important to regulate the attachment of contractile proteins to the plasma membrane of smooth muscle cells to allow force transmission to the extracellular matrix. For example, it has been suggested that activation of tyrosine kinases mediates the phosphorylation of the focal adhesion adapter protein–paxillin that then mediates contractile protein attachment to the plasma membrane and force transmission. The time course of smooth muscle energetics may reflect this complex mechanical linking. Typically, cross-bridge cycling in smooth muscle is much faster initially and then transitions to a slower "latch bridge" state of slow cycling rate. This transition to a latch bridge state may reflect to attachment of the contractile proteins to the plasma membrane, thereby increasing the load on cross-bridge cycling (see above).

In our airway smooth muscle, relaxation of force is probably more important than force generation. It is generally thought that relaxation occurs as a result of MyLC20 dephosphorylation. Obviously, a decrease in myoplasmic Ca^{2+} concentration in airway smooth muscle cells will lead to a decrease in MyLC20 phosphorylation. However, smooth muscle cells also contain myosin light-chain phosphatase (MLCP), which mediates dephosphorylation of the MyLC20 and thus, can regulate the extent of MyLC phosphorylation even in the presence of constant myoplasmic Ca^{2+} concentration. This is thought to be the major mechanism by which the sensitivity of force generation to myoplasmic Ca^{2+} concentration is regulated. Activation of myosin light-chain phosphatase is dependent on Rho kinase signaling. A number of agonists have been shown to activate the Rho kinase signaling pathway via G-protein-coupled receptors and thereby increase Ca^{2+} sensitivity of force generation in airway smooth muscles. For example, both ACh and histamine induce both an increase in myoplasmic Ca^{2+} concentration and an increase in Ca^{2+} sensitivity. Thus, force generation in airway smooth muscle reflects a balance between MLCK-mediated MyLC20 phosporylation, myosin light-chain phosphatase-mediated MyLC20 dephosphorylation (inhibited to increase Ca^{2+} sensitivity), and the activation of non-contractile, cytoskeletal remodeling.

Signaling pathways inducing smooth muscle relaxation are well studied. A major signaling pathway is mediated by nitric oxide, which in vascular smooth muscle is endothelium-derived. Nitric oxide exerts its effect by stimulating soluble guanylate cyclase that increases synthesis of cGMP. Other cell–cell interactions as well as neural and hormonal factors stimulate cAMP production via protein kinase G (PKG) and protein kinase A (PKA) activation. These cyclic nucleotides mediate phosphorylation of a number of proteins in airway smooth muscle cells that can lead to changes in myoplasmic Ca^{2+} concentration, MLCK activity and/or myosin light-chain phosphatase activity and as a result MyLC20 phosphorylation.

5.4   STRUCTURAL PROTEINS

Within the sarcomere, there are several proteins in addition to myosin and actin that compose both the sarcomere and the surrounding architecture of skeletal muscle. The Z-disc, which aligns the sarcomeres, is composed of α-actinin, a component of F-actin that provides structural support and a mechanism for cohesion and cell-to-cell interactions. Nebulin accounts for 2–3% of the protein content of muscle. It spans the thin filament and is anchored at the Z-disc. Nebulin is thought to regulate the length of the thin filament. Titin is a very large connecting protein within the sarcomere that maintains structural integrity of the sarcomere as well as providing mechanosensory feedback during contraction. Thus, titin may function as a stress–response protein that triggers downstream signaling pathways in skeletal muscle fibers.

There are various disorders of sarcomeric proteins including both muscular dystrophies and non-dystrophic myopathies. A common muscular dystrophy highlights the importance of structural proteins in sarcomeric architecture in skeletal muscle. Dystrophin is a structural protein that links the sarcomeric cytoskeleton to the extracellular matrix (see above). The lack of dystrophin is responsible for Duchenne muscular dystrophy, results in progressive muscle weakness, destabilizing of the sarcolemma, and susceptibility to contraction-induced injury.

5.5   MUSCLE FIBER TYPE CLASSIFICATION

For more than a century it has been recognized that different types of muscle fibers exist that vary considerably in the mechanical and fatigue properties. Initially, fibers were broadly classified as "red" versus "white" fibers, which we now know reflects the presence or absence of myoglobin, respectively. However, schemes for classifying different fiber types within skeletal muscle have progressed over the years. In one classification scheme, muscle fiber types were distinguished based on histological staining of metabolic enzyme activities. In this scheme, fibers were classified as slow oxidative (SO), fast oxidative, glycolytic (FOG) and fast, glycolytic (FG) [74, 106]. Another classification scheme was based on the pH lability of actomyosin ATPase staining in which fibers were classified as type I, IIa, IIb, or IIx. More recently, it was shown that MyHC isoform composition of muscle fibers generally corresponds with the classification scheme based on the pH lability of actomyosin ATPase staining [10, 109]. Identifying different MyHC isoforms was facilitated by the development of specific antibodies. Today, fiber type classification is primarily based on immunoreactivity to these specific MyHC isoform antibodies [87, 95]. These antibodies can also be used in Western blot analyses of MyHC isoform composition of single muscle fibers in which mechanical properties are determined. From these studies, it is clear that MyHC isoform composition of skeletal muscle fibers underlies differences in specific force, maximum shortening velocity, ATP consumption and fatigability. Overall, muscle fibers composed of the slow MyHC isoform (MyHC$_{slow}$—type I fibers) generate less specific force than fibers composed of fast MyHC isoforms [33, 34]. Among fast fibers, those containing the 2A isoform (MyHC$_{2A}$—type IIa fibers) generate less specific force than those containing the 2B (MyHC$_{2B}$—type IIb fibers) or 2X (MyHC$_{2X}$—type IIx fibers) isoforms. However, when normalized for MyHC content per half sarcomere (approximating force per MyHC molecule or cross-bridge assuming the same α_{fs}), there are no differences in force among fast fibers, but slow fibers still generate less force. This suggests that the average force per cross-bridge is lower in fibers expressing MyHC$_{slow}$ compared to those expressing fast MyHC isoforms (Figure 26).

The maximum unloaded shortening velocity of muscle fibers can be determined using a slack test in which muscle fibers are maximally activated at optimal sarcomere length and then rapidly shortened in steps of 5–15% of optimal sarcomere length [53]. As the slack in muscle length is taken up during shortening, the fibers are unloaded and no force is generated. The time delay before force is redeveloped is measured and the slope of the line relating slack length to time delay before force redevelopment approximates the maximum unloaded shortening velocity (V_0). Using this slack test, it has been shown that fibers expressing MyHC$_{slow}$ (type I fibers) have slower V_0 followed in rank order by fibers expressing MyHC$_{2A}$, MyHC$_{2X}$, and MyHC$_{2B}$ (type IIa, IIx, and IIb fibers, respectively).

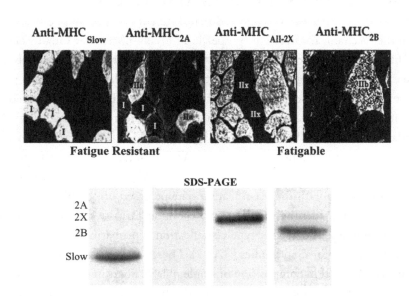

FIGURE 26: Diaphragm muscle fiber types can be identified based on immunoreactivity against antibodies specific for different MyHC isoforms. In addition, these antibodies can be used in Western blot analyses of single dissected muscle fibers. Used with permission from Sieck et al. [98].

In single dissected skeletal muscle fibers, cross-bridge cycling rate can be estimated using a test similar to the slack test, but in this case, after shortening the activated muscle fiber by ~20% of optimal length, the fiber is rapidly re-stretched to optimal length. Since during the rapid shortening, all cross-bridges are broken, force decreases to zero. As cross-bridges reattach, force redevelops, and the rate of this force redevelopment (k_{tr}) depends on cross-bridge cycling rate. As might be expected, fibers expressing $MyHC_{2X}$, and $MyHC_{2B}$ (type IIx and IIb fibers) have the fast k_{tr} followed in rank order by fibers expressing $MyHC_{2A}$, and $MyHC_{slow}$ (type IIa and I fibers) [102] (Figure 27).

The ATP consumption rate of single dissected skeletal muscle fibers can be measured using a stop flow technique in which ATP hydrolysis is coupled to the reduction of NADH (a fluorescent

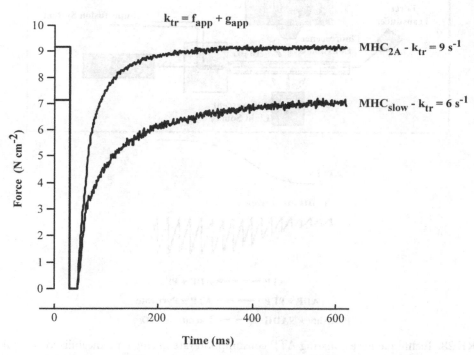

FIGURE 27: Different diaphragm muscle fiber types (distinguished by MyHC isoform expression) have varying cross-bridge cycling rates (k_{tr}). Cross-bridge cycling rate was determined in maximally activated (at pCa 4.0) single permeabilized diaphragm muscle fibers by quick release (shortening fiber length by 20% of optimal length—L_o) followed by rapid re-stretch back to L_o. With quick release, all cross-bridges are broken, and after re-stretch, as cross-bridges reattach and start to cycle, force redevelops. The rate constant for force redevelopment (k_{tr}) provides an estimate of cross-bridge cycling rate. In this example, a diaphragm muscle fiber expressing MHC_{2A} has a faster cross-bridge cycling rate compared to a fiber expressing MHC_{slow}.

molecule) to NAD (non-fluorescent) [41–43, 102]. Thus, the rate of extinction of NADH fluorescence directly relates to the rate of ATP consumption during force generation and shortening. Using this technique, it has been shown that fibers expressing $MyHC_{2X}$, and $MyHC_{2B}$ (type IIx and IIb fibers) have the highest ATP consumption rates followed in rank order by fibers expressing $MyHC_{2A}$ and $MyHC_{slow}$ (type IIa and I fibers) (Figure 28).

In addition to the four adult isoforms of MyHC, there are also embryonic ($MyHC_{emb}$) and neonatal ($MyHC_{neo}$) isoforms in skeletal muscle. During early myogenesis, primary myotubes ex-

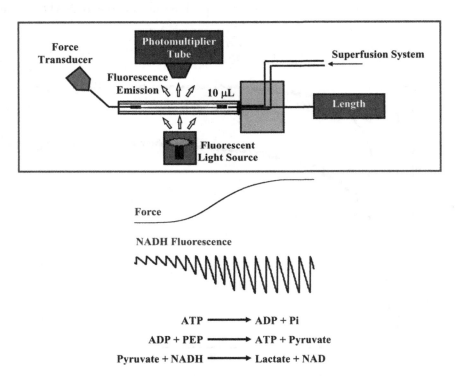

FIGURE 28: Technique for estimating ATP consumption rate in single permeabilized skeletal muscle fibers (or smooth muscle strips). Muscle fibers are mounted between a force transducer and a servomotor (for length control) in a quartz cuvette, into which a Ca^{2+} activating solution is perfused. The activating solution also includes an ATP regenerating system. When perfusion of the cuvette is stopped, the hydrolysis ATP during cross-bridge cycling is coupled to the reduction of NADH (a fluorescent molecule) to NAD (non-fluorescent) by phosphoenolpyruvate (PEP) and production of pyruvate. The rate of ATP consumption during force generation and shortening directly relates to the rate of extinction of NADH fluorescence.

press $MyHC_{slow}$ as well as $MyHC_{emb}$ isoforms. The progression from primary to secondary myotubes occurs as myotubes expressing the $MyHC_{emb}$ isoform transition into $MyHC_{neo}$. During late embryonic development, fibers expressing $MyHC_{slow}$, $MyHC_{emb}$, and $MyHC_{neo}$ can be found, sometimes with co-expression of these MyHC isoforms within a single fiber. The expression of $MyHC_{2A}$ emerges relatively early during postnatal development, whereas the expression of $MyHC_{2B}$ and $MyHC_{2X}$ appears later.

The co-expression of MyHC isoforms within single skeletal muscle fibers occurs under a variety of conditions. In fact, co-expression of $MyHC_{2B}$ and $MyHC_{2X}$ is quite relatively common even under normal conditions. In contrast, the co-expression of other MyHC isoforms in single muscle fibers is typically reflecting pathology or adaptation. For example, co-expression $MyHC_{slow}$ and $MyHC_{2A}$ and/or $MyHC_{2X}$ has been shown following conditions of muscle unloading. Co-expression of $MyHC_{slow}$ and $MyHC_{2A}$ and/or $MyHC_{2X}$ isoforms is also found in neurodegenerative diseases such as spinal muscular atrophy and amyotrophic lateral sclerosis. At the present time, it is unknown how MyHC isoform co-expression occurs. However, co-expression of MyHC isoforms obscures unambiguous classification of different muscle fiber types.

5.6 FIBER TYPE COMPOSITION OF RESPIRATORY MUSCLES

A number of studies have reported the fiber type composition of the diaphragm muscle in a variety of species. Generally, it has been observed that all fiber types are represented in the diaphragm muscle, but the proportions of each type may vary depending on species, age, and other conditions. As will be discussed below, the fiber type composition of respiratory muscles reflects the diversity of motor units that underlies neuromotor control and the range of motor behaviors that can be accomplished.

In the diaphragm muscle of most species, type IIx and IIb fibers have greater cross-sectional areas than type IIa and I fibers [64, 71, 100, 114]. There are also differences in maximum specific force across diaphragm muscle fiber type [31–34] with rank order of type IIb, type IIx, type IIa, and type I. Thus, differences in fiber type proportions do not reflect the relative contributions of the different fiber types (and motor unit types—see below) to the total force generated by a respiratory muscle.

. . . .

CHAPTER 6

Neural Control of Respiratory Muscles

Breathing is a motor pattern that eventually involves the coordinated activation of a number of our muscles, pump muscles and upper airway muscles (skeletal muscles), as well as smooth muscles of the trachea, bronchi, and lung airways. The neural pathways involved in the control of skeletal muscles differ from those involved in controlling airway smooth muscles. A major difference is the presence of a direct neural synapse on skeletal muscle fibers, the neuromuscular junction, and neuromuscular transmission that involves release of acetylcholine (ACh) and activation of nicotinic cholinergic receptors (ionotropic receptors). In contrast, airway smooth muscle cells do not have direct neural synapses or distinct neuromuscular junctions similar to those found on skeletal muscle fibers. A variety of neurotransmitters affect airway smooth muscle cells acting through both ionotropic and metabotropic receptors. Thus, the neural influence on airway smooth muscle cells reflects a balance of neural effects that are integrated with other agonist and antagonist effects.

For pump and upper airway respiratory muscles, the neural pathways are in many respects similar to those activating limb skeletal muscles. Most skeletal muscles are involved in distinct motor patterns such as the rhythmic patterns of breathing or walking. These motor patterns are organized in the central nervous system as a central pattern generator. The motor pattern itself can be influenced by a variety of sensory inputs as well as sleep–waking state, emotional state, circadian, ultradian, infradian rhythms, etc. The output of the central pattern generator is transmitted to premotor neurons that are typically located close to the central pattern generator. As with the central pattern generator, the excitability of premotor neurons can be influenced by a variety of inputs that help shape the overall descending premotor motor drive to motor neurons. Motor neurons represent the final output of the motor control system and are organized as motor units comprising a motor neuron and the group of muscle fibers it innervates. Motor neurons also receive a variety of excitatory and inhibitory afferent inputs that affect their excitability. Generally, the strength of the descending premotor drive will determine the strength of our inspiratory (or expiratory) efforts as reflected by tidal volume. The timing of inspiration versus expiration; thus, respiratory rate and duty cycle are determined by the central pattern generator and only transmitted by premotor neurons to motor neurons. It is now thought that the descending premotor input to respiratory motor neurons is widely distributed, and that the activation or recruitment of specific motor neurons (motor units—see below) depends on the intrinsic size-related electrophysiological properties of motor neurons. Thus, motor neurons are typically recruited in a specific size-related order—smallest to largest (see below)—during most motor behaviors. However, as mentioned above, the excitability of a motor neuron can also be affected by a variety of afferent inputs, including proprioceptive feedback, that convey information about the efficacy of force generation and contraction of muscle fibers. Motor neuron excitability is also affected by descending neuromodulator pathways that are both excitatory and inhibitory (Figure 29).

The most important outcome of our breathing is gas exchange in the lungs. Accordingly, there is sensory feedback relaying the efficacy of gas exchange via chemoreceptors. There are two general types of chemoreceptors; those sensitive to decreased O_2 levels (hypoxia) and those sensitive to increased CO_2 levels (hypercapnia). The chemoreceptors sensing hypoxia are located pe-

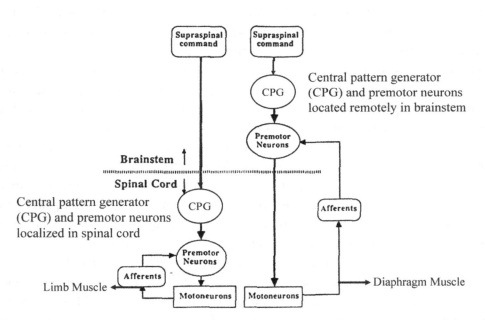

FIGURE 29: Neural control of respiratory muscles compared to limb muscles. Rhythmic motor patterns are generated within the central nervous system by a central pattern generator (CPG). The CPG for breathing is located in the brainstem, whereas the CPG for locomotor patterns is located in the spinal cord. In both cases, the output of the CPG is transmitted to premotor neurons, which for the phrenic system are located in the medulla, but for the locomotor system are located in the spinal cord in close proximity to the CPG. The excitatory drive output from premotor neurons is then diffusely transmitted to motor neurons. The recruitment of motor neurons (motor units) depends on their intrinsic, size-related electrophysiological properties.

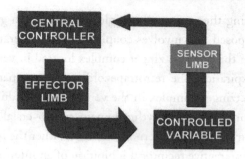

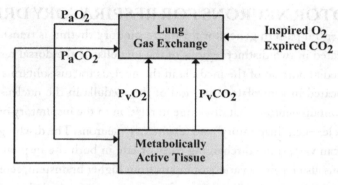

FIGURE 30: The control of respiration is an example of a biofeedback system. The efficacy of the respiratory motor system is sensed by chemo- and mechanoreceptors. Central and peripheral chemoreceptors relay sensory afferent feedback regarding the efficacy of gas exchange in the lung. Lung stretch receptors relay sensory afferent feedback regarding the efficacy of lung inflation.

ripherally, so they are called peripheral chemoreceptors. In contrast, the chemoreceptors sensing hypercapnia are located in the central nervous system in the brainstem and are thus called central chemoreceptors (Figure 30).

For gas exchange to occur, our lungs must inflate with fresh air. Accordingly, there is sensory feedback conveying information regarding the extent of our lung inflation. These sensory receptors are sensitive to the mechanical stretch of our lungs during inflation and are thus called lung stretch receptors.

6.1   CENTRAL PATTERN GENERATOR FOR BREATHING

The central pattern generator for breathing is best depicted as a two-phase model: 1) inspiration and 2) expiration; although a three-phase model is often presented with two distinct phases of expiration based on the activation of upper airway muscles. The rhythmic pattern of our breathing is classically viewed as reflecting the output of a single central pattern generator, although an alternative model has been proposed that involves coupled pattern generators that are anatomically distinct; one for inspiration in the pre-Boetzinger complex located in ventrolateral medulla of the brainstem, and second for expiration, the retrotrapezoid nucleus or parafacial respiratory group located rostral to the pre-Boetzinger complex in the ventrolateral medulla. The mechanisms underlying the generation of our respiratory rhythm remain controversial and the focus of intense investigation, but generally our breathing pattern results from either the intrinsic electrophysiology of pacemaker neurons or the interactive reciprocal inhibition of an interneuronal network.

6.2 PREMOTOR NEURONS FOR RESPIRATORY DRIVE

The output of the central pattern generator for the respiratory rhythm is transmitted to premotor neurons that are located in two distinct regions of the medulla: 1) the dorsal respiratory group located in the dorsomedial portion of the medulla in the nucleus tractus solitarius and 2) the ventral respiratory group located in ventrolateral portion of the medulla in the nucleus ambiguous. Both respiratory groups contain neurons that discharge in relation to the inspiratory or expiratory phases of the respiratory cycle—i.e., "inspiratory" and "expiratory" neurons. The discharge patterns of these medullary neurons can vary, some discharging early or late in both the inspiratory and expiratory phases. These neurons also receive a variety of inputs from higher brainstem, cerebellar, and cortical regions of the brain. An important distinction of premotor neurons is their monosynaptic projection to motor neurons innervating respiratory pump muscles (e.g., diaphragm and intercostal muscles) as well as upper airway muscles (e.g., genioglossus and posterior cricoarytenoid muscles). It is now fairly established that these premotor neurons exert an excitatory synaptic drive to respiratory motor neurons via glutamatergic neurotransmission. It is also clear that respiratory premotor neurons receive a variety of inputs that can shape the respiratory drive to respiratory motor neurons. For example, these medullary premotor neurons receive input from peripheral and central chemoreceptors as well as lung stretch receptors (see below).

6.3 MOTOR NEURONS INNERVATING RESPIRATORY MUSCLES

With the exception of a few upper airway muscles (e.g., genioglossus muscle), our respiratory muscles are innervated segmentally by motor neurons located in the cervical and thoracic spinal cord. Overall, there is a somatotopic organization in the location of motor neurons in the ventral horn of our spinal cord, with motor neurons innervating more distal limb muscles located dorsolaterally, whereas more proximal limb muscles are innervated by motor neurons located dorsomedially. In contrast, our trunk or axial muscles are innervated by motor neurons located ventromedially.

The phrenic motor neuron pool is situated in the ventromedial portion of the mid to lower cervical spinal cord, typically spanning three cervical segments. For example, in humans, phrenic motor neurons are located in spinal cord levels C3 through C5. The precise location of phrenic motor neurons can vary slightly across species. For example, in the cat, phrenic motor neurons are located in the C4 through C6 spinal cord segments. The location of phrenic motor neurons in the cervical spinal cord reflects the embryonic origin of the diaphragm muscle. One misconception is that the crural region of the diaphragm represents a separate muscle [21], with its own specific somatotopic pattern of innervation. However, it is now well established that all regions of the diaphragm muscle are innervated by phrenic motor neurons located in the cervical spinal cord.

As mentioned above, motor neurons innervating all intercostal muscles (external, internal, innermost intercostal muscles as well as subcostalis and transversus thoracis muscle) are located in the T1 through T11 spinal cord segments corresponding to the location of muscle fibers.

The abdominal muscles are also innervated segmentally by branches of the lower six intercostal nerves sometimes called the thoracoabdominal nerves with motor neurons located at spinal cord levels T7 through T11 and by the subcostal nerve (with motor neurons located at T12).

Motor neurons innervating upper airway muscles that affect placement of the hyoid bone (infrahyoid and suprahyoid muscles) are located in the upper cervical spinal cord from C1 through C3. The nerves that innervate the infrahyoid muscles are sometimes called the ansa cervicalis.

The genioglossus muscle together with other tongue muscles is innervated by the hypoglossal nerve (cranial nerve XII). Hypoglossal motor neurons are located in a column called the hypoglossal nucleus situated in the dorsomedial medulla near the floor of the 4th ventricle.

The recurrent laryngeal nerve innervates all intrinsic muscles of the larynx, including the posterior cricoarytenoid muscles that abduct the vocal folds during inspiration. The only exception is the cricothyroid muscle, which is innervated by the superior laryngeal nerve. Both the recurrent and superior laryngeal nerves are branches of the vagus nerve (cranial nerve X) with motor neurons located in the motor nucleus of the vagus in the dorsomedial medulla near the floor of the 4th ventricle and the hypoglossal nucleus.

6.4   MOTOR UNITS IN RESPIRATORY MUSCLES

Respiratory muscle motor units comprise a motor neuron and the group of muscle fibers that it innervates. Charles Sherrington first proposed the term motor unit in the early 20th century. He recognized that this motor neuron output from the central nervous system is the final common pathway for neural control of muscle contraction [66]. Much later, it was established that the electrophysiological properties of motor neurons and the mechanical and fatigue properties of muscle fibers comprising a motor unit are well matched. Thus, this matching of motor neuron electrophysiological and muscle fiber properties provides the foundation for neural control for a wide range of motor behaviors of skeletal muscles (Figure 31).

In most respiratory muscles, the mechanical and fatigue properties of muscle fibers vary considerably depending on their contractile protein composition and oxidative capacity; thus, providing the basis for diversity in the motor behaviors that can be accomplished. Based on these differences in the mechanical and fatigue properties of muscle fibers, motor units are classified into four different types [11, 12, 26, 96]. This classification of motor unit types mirrors the classification of muscle fiber types and reflects the fact that within a motor unit, fiber type composition is homogeneous [24, 96, 97]. Thus, similar to the classification of four muscle fiber types (see above), there are four types of motor units: 1) Type S motor units composed of type I muscle fibers that express the MyHC$_{slow}$ isoform and have higher oxidative capacities and mitochondrial volume densities; 2) Type FR motor units composed of type IIa fibers that express the MyHC$_{2A}$ isoform and also have higher oxidative capacities and mitochondrial volume densities; 3) Type FInt motor units composed of type IIx fibers that express the MyHC$_{2X}$ isoform and have intermediate oxidative capacities and mitochondrial volume densities; and 4) Type FF motor units composed of type IIb fibers that express predominantly the MyHC$_{2B}$ isoform but usually in combination with the MyHC$_{2X}$ isoform and have lower oxidative capacities and mitochondrial volume densities [24, 97].

Type S motor units display slower mechanical properties (slower twitch contraction times and maximum shortening velocities) [89]. These type S motor units are also more resistant to fatigue when repetitively activated; reflecting the slower cross-bridge cycling rates, lower ATP consumption rates, and higher oxidative capacities of their muscle fibers. Three motor unit types display faster mechanical properties (twitch contraction times and maximum shortening velocities) but have a range of susceptibilities to fatigue when repetitively stimulated that distinguishes them into different types [112]. Type FR motor units are "fast twitch" but as resistant to fatigue during repetitive stimulation as type S units. The resistance to fatigue of type FR motor units reflects the higher oxidative capacities and mitochondrial volume densities of type IIa fibers. Type FF motor units are also "fast-twitch," but they are highly fatigable during repetitive stimulation. The increased susceptibility to fatigue of type FF motor units reflects the faster cross-bridge cycling rates and ATP consumption rates of type IIb muscle fibers but also their lower capacities for oxidative phosphorylation and ATP production. Type FInt motor units are also "fast-twitch" but have an intermediate resistance to fatigue during repetitive stimulation. Again, the susceptibility of type IIx fibers is most likely a reflection of the balance between ATP consumption rate during cross-bridge cycling and the capacity for oxidative phosphorylation and ATP production [97].

In addition to differences in twitch contraction times, maximum shortening velocities and fatigue resistance, muscle fibers in different motor unit types also exhibit differences in other mechanical properties that reflect the properties of the muscle fibers comprising these units. For example, motor unit types have varying maximum specific forces with type S motor units displaying

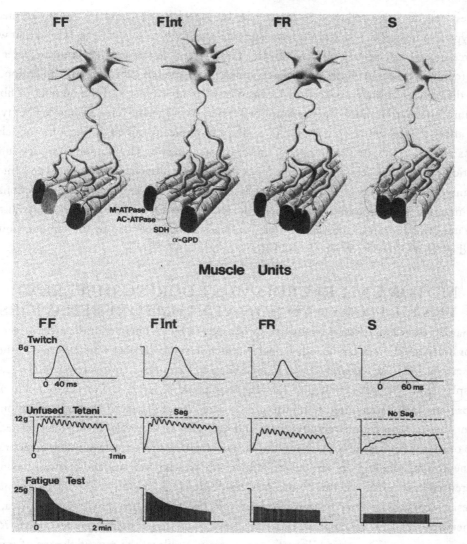

FIGURE 31: Classification of motor unit (muscle fiber) types. Four types of motor units are classified based on mechanical and fatigue properties of muscle fibers that depend on fiber type—slow-twitch, fatigue resistant (S), fast-twitch, fatigue resistant (FR), fast-twitch, fatigue intermediate (FInt), and fast-twitch, fatigable (FF). The mechanical properties used to distinguish motor unit types include twitch contraction time, presence of sag, and fatigue resistance. In this example, fiber types were classified based on the pH lability of myosin ATPase staining—M-ATPase staining higher for fast-twitch fibers. The oxidative capacity of fibers was assessed by determining succinate dehydrogenase (SDH) activity, whereas glycolytic capacity was determined by α-glycerophosphate dehydrogenase (α-GPD) activity. Recent studies employ difference in MyHC isoform expression to define fiber type. Used with permission from Sieck [89].

the lowest specific force followed in rank order by type FR, FInt and FF units. As mentioned above, type S motor unit muscle fibers have a greater sensitivity to myoplasmic $[Ca^{2+}]$ that will shift their force–frequency response curves leftward. Thus, type S motor units generate a greater level of relative (compared to maximum force) force at any given level of submaximal stimulation. There are also differences in the cross-sectional areas of muscle fiber types and the number of fibers in a motor unit (innervation ratio) that can amplify differences in specific force of muscle fiber types and thereby affect the total force contributed by a motor unit once recruited during a motor behavior.

As mentioned above, the number of muscle fibers innervated by a phrenic motor neuron (i.e., innervation ratio) varies across motor unit types with greater innervation ratios at type FInt and FF motor units than at type S or FR units [89]. Taken together, the larger fiber cross-sectional areas, greater innervation ratios and greater specific forces of type FInt and FF motor units in the diaphragm muscle, all result in substantially greater forces being generated by these motor units when recruited compared to that of type S and FR motor units [67, 89].

6.5   MOTOR UNIT RECRUITMENT DURING DIFFERENT VENTILATORY AND NON-VENTILATORY BEHAVIORS

More than 50 years ago, Elwood Henneman proposed the Size Principle for the orderly recruitment of motor units based on differences in axonal conduction velocities that were reflected in the size of motor neurons [45]. As recognized by Henneman, smaller motor neurons have smaller axons and thus display slower axonal conduction velocities compared to larger motor neurons. He observed that those motor units recruited first during motor behaviors displayed slower conduction velocities compared to motor units recruited later and at higher force levels. Henneman related this to the size of motor neurons and the impact of size on their intrinsic electrophysiological properties. Motor neuron morphology, in particular dendritic arborization and somal dimensions, contributes to differences in intrinsic electrophysiological properties (e.g., membrane capacitance and membrane resistance) across motor neurons. For example, because of their smaller surface area, smaller motor neurons have lower membrane capacitance (C_m) and higher membrane resistance (R_m) than larger motor neurons. The excitability of a motor neuron reflects the rate of change of membrane potential (dV_m/dt) for a given amount of input current (I_c or capacitive current). Mathematically, this can be expressed by the following equation:

$$dV_m/dt = I_c/C_m$$

Thus, for a given level of input I_c, smaller motor neurons, with lower C_m, have a greater change in V_m are thus more excitable. Another way to look at this is that to reach a threshold for generation of an action potential, smaller motor neurons requires less input current than larger motor neurons. The amount of input or synaptic current required to generate an action potential in a neuron is referred to as the rheobase. Thus, smaller motor neurons have a lower rheobase than larger motor neurons. It is now recognized that motor neurons innervating type S and FR motor units are smaller and thus more excitable (i.e., lower rheobase) than motor neurons innervating type FInt and FF motor units [11, 113].

There is considerable morphological heterogeneity in motor neuron size, even within a single pool of motor neurons [13, 15, 69, 79]. For respiratory motor neurons, such as phrenic motor neurons that receive distributed excitatory synaptic input (descending respiratory drive from medullary premotor neurons), smaller more excitable motor neurons with smaller axons and slower conduction velocities are recruited before larger motor neurons. Indeed, there is substantial evidence that this orderly recruitment of motor units exists in the diaphragm muscle [22, 50, 51, 93], consistent with the size principle. For example, studies have shown that phrenic motor neurons with slower axonal conduction velocities are recruited first during inspiratory efforts [22, 50, 51]. Accordingly, recruitment of diaphragm motor units generally matches the mechanical and fatigue properties of their muscle fibers: type S and FR motor units are recruited first, followed by type FInt and FF units.

A model for diaphragm motor unit recruitment across a range of ventilatory and non-ventilatory behaviors has been proposed based on the force generated per motor unit type and the assumption of an orderly recruitment in rank order type S, FR, FInt, and FF. Since it is difficult to measure the force of a single motor unit, this can be estimated based on: 1) measurements of the specific force (force per cross-sectional area) of single type identified diaphragm muscle fibers, 2) measurements of the proportion and cross-sectional areas of different fiber types in the diaphragm muscle, 3) an estimate of the number of phrenic motor neurons, and 4) an assumption that differences in innervation ratio (i.e., the number of fibers innervated by each motor neuron) of different motor unit types measured in the cat holds true for other species (i.e., innervation ratio of type S or FR motor units is ~15% lower than that for type FInt and FF units) [24, 94, 96, 97]. The relative proportion of fatigue-resistant type S and FR motor units versus more fatigable type FInt and FF motor units varies across species. For example, in the rat diaphragm muscle, the combined proportion of type S and FR motor units is ~65% compared to only ~34% in the cat [88, 93]. As mentioned above, the force contributed by diaphragm motor units types varies considerably (type FF > FInt > FR > S). The proportionally larger force generated by type FF and FInt motor units results from a combination of factors, greater specific force of type IIb and IIx fibers, larger cross-sectional areas of type IIx and IIb fibers, and increased innervation ratios. Although type FInt and FF motor units contribute greater force this is at the expense of increased susceptibility to fatigue if repeatedly activated. This is an important consideration for the diaphragm muscle where the duty cycle for inspiratory activation is ~40%. At such an activation duty cycle, type FInt and FF motor units would rapidly fatigue (Figure 32).

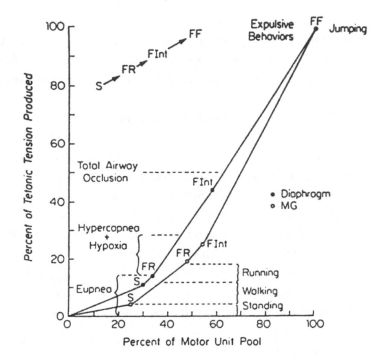

FIGURE 32: Model of motor unit recruitment in the cat diaphragm and medial gastrocnemius muscles. The model assumes an orderly recruitment of motor units in the following rank order—type S, FR, FInt and FF. In this case, the model also assumed maximum activation of one motor unit type before recruitment of the next type. Note that in both cases, sustained motor behaviors (ventilation for the diaphragm, standing and walking for the medial gastrocnemius) can be accomplished by the recruitment of only fatigue resistant type S and FR motor units. Used with permission from Sieck and Fournier [92].

The model for motor unit recruitment can also assume unit activation at different frequencies. Motor unit discharge frequencies may change during inspiration [14, 59, 107]. In previous studies, it has been shown that the discharge frequency of diaphragm motor units at the onset of activity is lower than at times later within an inspiratory burst [49, 107]. In the cat diaphragm muscle, motor units recruited earlier during inspiration had onset discharge frequencies of ~15 Hz and peak discharge rate of ~25 Hz. In contrast, motor units recruited later during inspiration had onset discharge frequencies of ~20 Hz and peak discharge rates of ~40 Hz. This range of discharge frequencies approximately covers the steep portion of the force frequency response curve for type S and FR motor units.

In a number of studies, it has been shown that there is a strong correlation between the Pdi generated by the diaphragm muscle and electromyographic (EMG) activity as quantified by calculating peak root-mean-square (RMS) [67, 91, 110]. The RMS EMG amplitude increases proportionately with changes in Pdi, and thus, diaphragm muscle EMG activity can be used as a surrogate for Pdi measurements, particularly when these are impractical (e.g., in chronic measurements or in awake humans or animals).

In previous studies in cats [26, 93], in hamsters [90], and more recently, in rats [67], Pdi and EMG measurements were used to estimate the forces generated by the diaphragm muscle during different ventilatory and non-ventilatory behaviors. For comparison, maximum Pdi (Pdimax) was measured in response to bilateral phrenic nerve stimulation. In these studies, it was observed that the Pdi generated during quiet breathing (eupnea) varies across species; in cats, it is ~12% of Pdimax, in hamsters and rats, ~27% and ~21% of Pdimax, respectively, and in humans, ~10% of Pdimax [90]. Based on the model of diaphragm motor unit recruitment described above, the Pdi generated during eupnea can be accomplished by the recruitment of on fatigue resistant type S and FR motor units in all species [67, 88, 90, 93].

When ventilation is stimulated by hypoxia (10% O2) and hypercapnia (5% CO2), Pdi increases to ~28–30% of Pdimax in cats and rats [67, 93]. Thus, while hypoxia–hypercapnia exposure provides a very powerful ventilatory stimulus, it does not maximally drive diaphragm muscle force generation or the recruitment of all diaphragm motor units. In fact, based on the motor unit recruitment model, this level of Pdi can be accomplished by the recruitment of only type S and FR motor units in the diaphragm muscle. In agreement with this estimate of motor unit recruitment, it has been estimated that the number of phrenic motor neurons in cats that display spontaneous inspiratory-related discharge is ~23% of the total motor neuron pool [50]. This is the approximately the proportion of type S and FR motor units in the cat diaphragm muscle [26, 93].

During sustained airway occlusion, the Pdi generated increases to ~50% of Pdimax in cats, and ~63% of Pdimax in rats [67, 88, 90, 93]. Based on the recruitment model, this level of Pdi can be achieved by the full recruitment of all type S and FR motor units with the additional recruitment of some FInt motor units.

Only during non-ventilatory expulsive behaviors of the diaphragm muscle (e.g., coughing, sneezing) does the Pdi generated by the diaphragm muscle approximate Pdimax in both cats and rats. Thus, during expulsive non-ventilatory motor behaviors, full recruitment of all diaphragm motor unit types is required. However, these efforts are relatively infrequent, so motor unit fatigue with sustained activity would not be a complicating factor in generating sufficient Pdi to accomplish these important behaviors.

Ventilatory and non-ventilatory demands for Pdi generation vary across species, and are matched by the relative proportions of fatigue-resistant type S and FR motor units versus more fatigable type FInt and FF motor units (see above). The relative Pdi generated during ventilatory behaviors reflects the reserve capacity of the diaphragm muscle to accomplish its major ventilatory function compared to other non-ventilatory demands. It is likely that in larger species, there is a greater requirement for expulsive behaviors involved in clearing the airways. In addition, differences in the extent of diaphragm muscle activation across species may be related to the mechanical properties of the lung or chest wall and airway resistance.

These results in the diaphragm muscle verify considerable force reserve to generate the Pdi necessary to sustain ventilation. This force reserve reflects the population of motor units within the diaphragm muscle. Changes in fiber type composition or cross-sectional area of fibers in respiratory muscles following injury or disease will impact the ability of these muscles to generate forces required for ventilation placing patients at increased risk for respiratory failure [101]. In addition, the ability of the diaphragm muscle to generate more strenuous Pdi to accomplish expulsive non-ventilatory behaviors may be compromised, also increasing the risk of patients for pulmonary diseases associated with inadequate clearance of the airways, e.g., pneumonia.

6.6   NEUROMUSCULAR TRANSMISSION

The neuromuscular junction is the synapse between a motor neuron and the muscle fiber it innervates. Within a motor unit, there are as many neuromuscular junctions as there are muscle fibers comprising the unit, but at each muscle fiber, there is a single neuromuscular junction, and it is the final effector for neural control of contraction of that muscle fiber. Motor units display considerable diversity in terms of the electrophysiological properties across motor neurons and the mechanical and fatigue properties of muscle fibers. A hallmark of neuromotor control is the fact that the properties of motor neurons and muscle fibers are matched to affect appropriate recruitment of specific types of motor units to accomplish a variety of motor functions. As the point of contact between a motor neuron and muscle fiber, the neuromuscular junction must also be able to meet various functional demands via neuromuscular transmission. Therefore, it is not surprising that the structural and functional properties of neuromuscular junctions vary with motor unit/muscle fiber type.

The morphology of neuromuscular junctions at different muscle fiber types is quite distinct [80]. For example, neuromuscular junctions at types I and IIa diaphragm muscle fibers (type S and FR motor units, respectively) are smaller and display less complexity in branching compared to neuromuscular junctions at types IIx and/or IIb fibers (type FInt and FF motor units, respectively) [76, 104]. However, these differences in neuromuscular junction morphology may not reflect fiber type *per se*. In the soleus muscle, which is predominantly composed of type I muscle fibers, neuromuscular junctions are larger compared to neuromuscular junctions in the extensor digitorum longus muscle, which is predominantly composed of type IIx and/or IIb fibers. Perhaps, it is fiber cross-sectional area that is the major determinant of the size of neuromuscular junctions regardless of fiber type. Other factors such as age, species, and activation history may also play important roles in determining the morphological properties of neuromuscular junctions [81, 83] (Figure 33).

Differences in the ultrastructure of pre- and postsynaptic elements of neuromuscular junctions also exist across motor unit/muscle fiber types. For example, the surface areas of presynaptic terminals at type IIx and/or IIb fibers in the diaphragm muscle are larger compared to presynaptic terminals at type I and IIa fibers. At the presynaptic terminal, the active zone is the point at which synaptic vesicles are released to affect neurotransmitter (ACh) mediated neuromuscular transmission (see below). While the density of active zones at the surface of presynaptic terminals is similar across muscle fiber types, the fact that there are differences in total surface area means that the total number of active zones is greater at type IIx and/or IIb fibers compared to type I and IIa fibers. At the active zone, some synaptic vesicles are "docked" or fused to the terminal membrane and are thus ready for immediate release. The number of docked synaptic vesicles at each active zone is comparable across different fiber types. Yet, since there are a greater number of active zones at type IIx and/or IIb fibers, the total number of docked, readily-releasable synaptic vesicles is greater at these fibers compared to type I and IIa fibers. Thus, in response to a single stimulus pulse, more synaptic vesicles are released at type IIx and/or IIb fibers—i.e., quantal content is higher at these fibers.

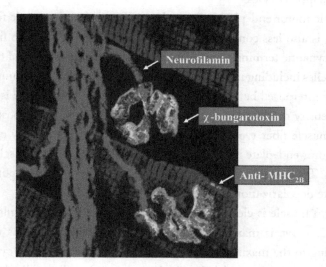

FIGURE 33: Confocal fluorescent image of diaphragm muscle neuromuscular junctions. Single type IIb muscle fibers are labeled with anti-MyHC$_{2B}$ (blue). Motor end plates are labeled with α-bungarotoxin (green), and axon terminals are labeled with an anti-neurofilamin antibody (red).

At presynaptic terminals, a large number of synaptic vesicles are not docked at active zones; but constitute a reserve pool that can be recruited to sustain synaptic vesicle release with repetitive stimulation. The number of synaptic vesicles in this reserve pool is greater at presynaptic terminals at type I and IIa fibers compared to type IIx and/or IIb fibers. Presynaptic terminals at type I and IIa fibers also have greater mitochondrial volume density compared to type IIx and/or IIb fibers; thus, these presynaptic terminals at type I and IIa fibers are more able to meet the higher metabolic requirements of more frequent activation. Accordingly, the ultrastructure of presynaptic terminals at neuromuscular junctions innervating different motor unit/muscle fiber types appear to be specialized to meet the functional demands of the motor unit. Types IIx and IIb fibers are larger; thus greater synaptic drive (current to generate an excitatory postsynaptic potential—EPP) is required to reach threshold for action potential generation (see below). This functional demand is met by the fact that there are more synaptic vesicles readily available for release at the presynaptic terminals innervating type IIx and IIb fibers. However, this functional demand can only be met for a short time since there are fewer synaptic vesicles in reserve pools. In contrast, type S and FR motor units are activated more frequently; thus, the presynaptic terminals innervating type I and IIa muscle fibers must be able to sustain synaptic vesicle release with repetitive stimulation. This is accomplished by the presence of a larger reserve pool of synaptic vesicles at the presynaptic terminals of types I and IIa muscle fibers. In addition, synaptic vesicle recycling after release is effective in replenishing a releasable pool of synaptic vesicles.

At postsynaptic motor end-plates of type I and IIa diaphragm muscle fibers, branching and postsynaptic folding is also less complex compared to that at IIx and/or IIb fibers—matching the ultrastructure of presynaptic terminals. In addition, at the motor end-plates of type I and IIa muscle fibers, cellular organelles including mitochondria, rough endoplasmic reticulum, free polysomes, and nuclei are frequently interposed between the endplate and myofibrils. There is no evidence for any differences in the density of nicotinic cholinergic (ACh) receptors at the postsynaptic membrane across motor unit/muscle fiber types. However, it appears that the density of voltage-gated Na^+ channels near the motor end-plate is much higher at type IIx and/or IIb muscle fibers compared to type I and IIa fibers. This would effectively lower the threshold for action potential generation in response to end-plate depolarization (see below).

The diaphragm muscle is clearly susceptible to neuromuscular transmission failure. For example, if the phrenic nerve is maximally stimulated at ~40 Hz in 330 ms trains repeated each second (corresponding to the maximal motor unit discharge rates and duty cycle occurring during quiet breathing), the force generated by the diaphragm muscle will rapidly decline (e.g., an ~70% decrease in force within 2 minutes). However, if the diaphragm muscle is stimulated directly (bypassing the neuromuscular junction), the evoked force is much higher. The difference in diaphragm muscle force generated by phrenic nerve stimulation versus direct muscle stimulation reflects the extent of neuromuscular transmission failure [60, 108]. Under such non-physiological conditions, diaphragm muscle fatigue induced by repetitive phrenic nerve stimulation is predominantly due to neuromuscular transmission failure [103]. However, as mentioned above, to accomplish ventilatory behaviors of the diaphragm muscle, it is necessary to recruit only type S and FR motor units (type I and IIa muscle fibers). The neuromuscular junctions at type I and IIa fibers are specifically suited to sustain repetitive activation without failure. Type FInt and FF motor units in the diaphragm muscle are infrequently recruited and only for shorter durations. Thus, neuromuscular junctions at type IIx and IIb fibers are specialized to ensure reliable activation of these larger fibers for short periods of time [52]. However, they are not designed to sustain repeated activation over longer periods (e.g., as occurs during breathing) (Figure 34).

It has been clearly demonstrated that the efficacy of neuromuscular transmission varies across motor unit types in the diaphragm muscle, especially during repeated activation. The safety factor for neuromuscular transmission is determined by the ratio of end-plate potential (EPP) amplitude to the threshold for muscle fiber action potential generation. The amplitude of EPPs is greater at type IIx and/or IIb diaphragm muscle fibers, reflecting the higher quantal release of ACh. Recall from the equation above ($dV_m/dt = I_c/C_m$), the change in membrane potential (EPP) will depend directly on synaptic drive (I_c) as determined by quantal release but indirectly on muscle fiber membrane surface area (C_m). The membrane surface areas of type IIx and/or IIb diaphragm muscle fibers are 2–3 times larger than those at type I and/or IIa fibers. For action potential generation, this is partially offset by the higher density of voltage-gated Na$^+$ channels that effectively lower the

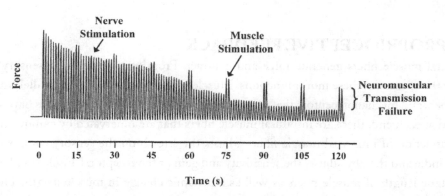

FIGURE 34: Neuromuscular transmission failure of the diaphragm muscle. The phrenic nerve was stimulated at 40 Hz in 300 ms trains repeated every s. In addition, direct muscle stimulation was superimposed every 15 s. Neuromuscular transmission failure was estimated by the difference between force induced by nerve stimulation versus direct muscle stimulation. Modified with permission from Sieck et al. [108].

threshold for action potential generation at type IIx and/or IIb fibers. Thus, overall, the safety factor for neuromuscular transmission during a single evoked response is higher at type IIx and/or IIb fibers compared to type I or IIa fibers. However, during repetitive stimulation, the EPP amplitude progressively declines across all fiber types, but the decline is much greater at type IIx and/or IIb fibers. Accordingly, type IIx and/or IIb diaphragm muscle fibers are much more susceptible to neuromuscular transmission failure.

Synaptic plasticity is a hallmark of the ability of the nervous system to adapt in response to both intrinsic and extrinsic stimuli, and forms the basis for changes in the efficacy of synaptic transmission. There are a number of changes that could affect synaptic plasticity at the neuromuscular junction: 1) neuromuscular activity may change due to changes in external load or altered ventilatory or non-ventilatory requirements; 2) muscle fiber size may change—atrophy or hypertrophy, or fiber type composition may be altered; and 3) the number of motor neurons may be reduced due to disease or trauma resulting in denervation of some muscle fibers, which are subsequently re-innervated by the sprouting of axons from remaining motor neurons [58, 68].

In the late 1940s, Donald Hebb introduced the concept of activity-dependent synaptic plasticity, which addresses the structural and functional changes that occur at synapses in response to altered use (either an increase or decrease in activity) [44]. According to the Hebbian theory, synaptic efficacy is enhanced when the extent of correlation between pre- and post-synaptic activity increases. With an increase or decrease in neuromuscular activity, synaptic plasticity may occur, but only if there are alterations in the fidelity of neuromuscular transmission (i.e., extent of correlation between pre- and postsynaptic activity) [70].

6.7   PROPRIOCEPTIVE FEEDBACK

Our skeletal muscle fibers generate force and contract. Proprioceptors provide sensory feedback regarding the efficacy of these motor functions. Muscle spindles are situated in parallel with skeletal muscle fibers and mechanoreceptors that sense muscle fiber length. Muscle spindles have a fusiform shape, and at each end, there are intrafusal muscle fibers that are innervated by gamma motor neurons. Contraction of intrafusal muscle fibers adjusts the stretch on the sensory component of the muscle spindle and thereby adjusts the sensitivity and gain of the receptor. Muscle spindles respond to both static length of muscle fibers as well as a dynamic change in muscle length. The afferent sensory feedback from muscle spindles synapse directly on the motor neurons that innervate the muscle in which they are located and exert an excitatory effect. At the same time, muscle spindle sensory feedback exerts an inhibitory effect on motor neurons innervating antagonist muscles via a disynaptic pathway. Thus, if a muscle spindle is located in a flexor muscle such as the biceps, its afferent sensory feedback will result in excitation to motor neurons innervating the biceps while exerting an inhibitory effect on motor neurons innervating the triceps muscle.

Golgi tendon organs are mechanoreceptors situated in series with muscle fibers in the tendon. As muscle fibers generate force Golgi tendon organs are activated; thus these receptors sense the efficacy of force generation. Afferent sensory feedback from Golgi tendon organs exerts an inhibitory effect on the motor neurons innervating the same muscle. Thus, Golgi tendon organs exert a simple negative feedback effect.

The diaphragm muscle has very few if any muscle spindles. In contrast, the intercostal muscles are richly supplied with muscle spindles, which appear to play an important role in postural adjustments of the ribcage. In addition, sensory feedback from muscle spindles located in the intercostal muscles can exert an excitatory effect on phrenic motor neurons. This effect has been termed the intercostal to phrenic reflex.

The central tendon of the diaphragm muscle contains Golgi tendon organs that respond to force generation by diaphragm muscle fibers. The precise role of Golgi tendon organs in modulating diaphragm force generation during ventilatory behaviors is unclear. However, sensory feedback from Golgi tendon may become important in limiting force generation during higher force, nonventilatory behaviors or the diaphragm muscle.

6.8   PERIPHERAL CHEMORECEPTORS

Peripheral chemoreceptors are located outside the central nervous system in specialized structures called the carotid and aortic bodies. The peripheral chemoreceptors respond to changes in the O_2 and CO_2 levels in the arterial blood. The carotid body is located in the neck at the division of the common carotid artery into the external and internal carotid arteries. The aortic body is located at the arch of the aorta. Both the carotid and aortic bodies contain sensory receptors that transmit afferent sensory nerve information back to the central nervous system. Afferent sensory nerves from the carotid body are located in the glossopharyngeal nerve (cranial nerve IX), whereas afferent sensory nerves from the aortic body are located in the vagus nerve (cranial nerve X).

Normally, the partial pressure of O_2 (pO_2) in the arterial blood is ~80 mm Hg and the partial pressure of CO_2 (pCO_2) is ~40 mm Hg. If arterial pO_2 decreases (hypoxia), the peripheral chemoreceptors are stimulated, which then triggers an increase in ventilation via an increase in respiratory rate or tidal volume. A change in arterial pCO_2 is normally sensed by the central chemoreceptors (see below); however, the peripheral chemoreceptors are also sensitive to an increase in pCO_2 and the resulting change in pH (see below). The sensitivity of the peripheral chemoreceptors to changes in pO_2 (and pCO_2) can be modified by neural efferent mechanisms. For example, there are efferent nerves to the carotid body that release dopamine, which acts as an inhibitory neurotransmitter to suppress the activity of peripheral chemoreceptors. Thus, the sensitivity of peripheral chemoreceptors can be acutely modulated. The sensitivity of peripheral chemoreceptors can also be affected by aging and chronic disease. For example, individuals with obstructive sleep apnea display repeated episodes of hypoxia. As a result, their peripheral chemoreceptor sensitivity to hypoxia is blunted.

The effect of peripheral chemoreceptor response to hypoxia is very evident when you are exposed to high altitudes, where the O_2 level in the ambient air is reduced. At high altitude, the fractional concentration of O_2 in the air remains the same at ~21%; however, the atmospheric pressure is reduced, thus the pO_2 is reduced. For example, at sea level, the atmospheric pressure is 760 mm Hg, which results in a pO_2 of ~160 mm Hg in the air. In contrast at 20,000 ft, the atmospheric pressure is 365 mmHg and the pO_2 is only 77 mm Hg. As a result, there is less O_2 available for alveolar gas exchange and hemoglobin in the arterial blood becomes less saturated with O_2. In response, the peripheral chemoreceptors are stimulated and ventilation increases.

6.9   CENTRAL CHEMORECEPTORS

As their name implies, central chemoreceptors are located within the central nervous system, specifically in the floor of the fourth ventricle of the brain stem. These are neurons that respond to changes in the pH (H^+ ion concentration) of the cerebrospinal fluid (CSF) as a result of changes in the level of CO_2 in the blood, which rapidly diffuses into the CSF. Through the carbonic anhydrase reaction, CO_2 levels in the CSF equilibrate with H^+ and HCO_3^- ion concentrations.

$$CO_2 + H_2O \xrightarrow{\text{Carbonic Anhydrase}} H^+ + HCO_3^-$$

If as a result of exercise or some other activity, blood CO_2 levels increase then CO_2 levels in the CSF increase pushing the reaction to the right; thereby increasing H^+ ion concentration (decreasing pH) and HCO_3^- ion concentration in the CSF. Normally, pH of the CSF is ~7.4. Under alkaline conditions, pH is > 7.4, and under acidic conditions, pH < 7.4. When the CSF pH is acidic, as during exercise, the central chemoreceptors respond to increase ventilation and thereby decrease CO_2 levels in the blood—a negative feedback mechanism. Conversely, if you hyperventilate (e.g., at high altitude in the example shown above), CO_2 levels in the blood will fall below normal and this will have an effect to inhibit ventilation.

· · · ·

References

[1]   Agur AMR and Dalley AF. Grant's Atlas of Anatomy. Baltimore: Lippincott Williams & Wilkins, 2009.

[2]   Allan DW, and Greer JJ. Embryogenesis of the phrenic nerve and diaphragm in the fetal rat. *J Comp Neurol* 382: pp. 459–68, 1997.

[3]   Ay B, Prakash YS, Pabelick CM, and Sieck GC. Store-operated Ca2+ entry in porcine airway smooth muscle. *Am J Physiol Lung Cell Mol Physiol* 286: pp. L909–17, 2004.

[4]   Babiuk RP, Zhang W, Clugston R, Allan DW, and Greer JJ. Embryological origins and development of the rat diaphragm. *J Comp Neurol* 455: pp. 477–87, 2003.

[5]   Bai TR, Bates JH, Brusasco V, Camoretti-Mercado B, Chitano P, Deng LH, Dowell M, Fabry B, Ford LE, Fredberg JJ, Gerthoffer WT, Gilbert SH, Gunst SJ, Hai CM, Halayko AJ, Hirst SJ, James AL, Janssen LJ, Jones KA, King GG, Lakser OJ, Lambert RK, Lauzon AM, Lutchen KR, Maksym GN, Meiss RA, Mijailovich SM, Mitchell HW, Mitchell RW, Mitzner W, Murphy TM, Pare PD, Schellenberg RR, Seow CY, Sieck GC, Smith PG, Smolensky AV, Solway J, Stephens NL, Stewart AG, Tang DD, and Wang L. On the terminology for describing the length-force relationship and its changes in airway smooth muscle. *J Appl Physiol* 97: pp. 2029–34, 2004.

[6]   Beachey W. *Respiratory Care Anatomy and Physiology*. St. Louis: Mosby, 1998.

[7]   Berne R, Levy M, Koeppen B, and Stanton B. Physiology. St Louis: Mosby, 2004.

[8]   Brenner B. Kinetics of the crossbridge cycle derived from measurements of force, rate of force development and isometric ATPase. *J Muscle Res Cell Motil* 7: pp. 75–6, 1986.

[9]   Brenner B, and Eisenberg E. Rate of force generation in muscle: Correlation with actomyosin ATPase activity in solution. *Proc Nat Acad Sci USA* 83: pp. 3542–6, 1986.

[10]   Brooke MH, and Kaiser KK. Three "myosin adenosine triphosphatase" systems: The nature of their pH lability and sulfhydryl dependence. *J Histochem Cytochem* 18: 670–2, 1970.

[11]   Burke RE. Motor units: anatomy, physiology and functional organization. In: *Handbook of Physiology. The Nervous System. Motor Control*, edited by Peachey LD. Bethesda, MD: Am Physiol Soc, 1981, pp. 345–422.

[12]   Burke RE, Levine DN, Tsairis P, and Zajac FE, 3rd. Physiological types and histochemical profiles in motor units of the cat gastrocnemius. *J Physiol (Lond)* 234: pp. 723–48, 1973.

[13]   Burke RE, Marks WB, and Ulfhake B. A parsimonious description of motoneuron dendritic morphology using computer simulation. *J Neurosci* 12: pp. 2403–16, 1992.

[14]   Butler JE, McKenzie DK, and Gandevia SC. Discharge properties and recruitment of human diaphragmatic motor units during voluntary inspiratory tasks. *J Physiol* 518(Pt 3): pp. 907–20, 1999.

[15]   Cameron WE, Averill DB, and Berger AJ. Quantitative analysis of the dendrites of cat phrenic motoneurons stained intracellularly with horseradish peroxidase. *J Comp Neurol* 230: pp. 91–101, 1985.

[16]   De Troyer A. Effect of hyperinflation on the diaphragm. *Eur Respir J* 10: pp. 708–13, 1997.

[17]   De Troyer A, and Estenne M. Coordination between rib cage muscles and diaphragm during quiet breathing in humans. *J Appl Physiol* 57: pp. 899–906, 1984.

[18]   De Troyer A, and Farkas GA. Contribution of the rib cage inspiratory muscles to breathing in baboons. *Respir Physiol* 97: pp. 135–46, 1994.

[19]   De Troyer A, Kirkwood PA, and Wilson TA. Respiratory action of the intercostal muscles. *Physiol Rev* 85: pp. 717–56, 2005.

[20]   De Troyer A, Sampson M, Sigrist S, and Macklem PT. Action of costal and crural parts of the diaphragm on the rib cage in dog. *J Appl Physiol* 53(1): pp. 30–9, 1982.

[21]   De Troyer A, Sampson M, Sigrist S, and Macklem PT. The diaphragm: two muscles. *Science* 213: pp. 237–8, 1981.

[22]   Dick TE, Kong FJ, and Berger AJ. Recruitment order of diaphragmatic motor units obeys Hennemans's size principle. In: *Respiratory Muscles and Their Neuromotor Control*, edited by Sieck GC, Gandevia SC, and Cameron WE. New York: Alan R. Liss, 1987, pp. 249–61.

[23]   DiMarco AF. Neural prostheses in the respiratory system. *J Rehabil Res Dev* 38: pp. 601–7, 2001.

[24]   Enad JG, Fournier M, and Sieck GC. Oxidative capacity and capillary density of diaphragm motor units. *J Appl Physiol* 67: pp. 620–7, 1989.

[25]   Fenn WO. A quantitative comparison between the energy liberated and the work performed by the isolated sartorius muscle of the frog. *J Physiol (London)* 58: pp. 175–203, 1923.

[26]   Fournier M, and Sieck GC. Mechanical properties of muscle units in the cat diaphragm. *J Neurophysiol* 59: pp. 1055–66, 1988.

[27]   Fournier M, and Sieck GC. Somatotopy in the segmental innervation of the cat diaphragm. *J Appl Physiol* 64: pp. 291–8, 1988.

[28]   Fredberg JJ, Jones KA, Nathan M, Raboudi S, Prakash YS, Shore SA, Butler JP, and Sieck GC. Friction in airway smooth muscle: Mechanisms, latch, and implications in asthma. *J Appl Physiol* 81: pp. 2703–12, 1996.

[29]   Fritsch H, and Kuehnel W. *Color Atlas of Human Anatomy*. New York: Thieme, 2008.

[30]   Gandevia SC, McKenzie DK, and Plassman BL. Activation of human respiratory muscles during different voluntary manoeuvres. *J Physiol* 428: pp. 387–403, 1990.

[31]   Geiger PC, Cody MJ, Han YS, Hunter LW, Zhan WZ, and Sieck GC. Effects of hypothyroidism on maximum specific force in rat diaphragm muscle fibers. *J Appl Physiol* 92: pp. 1506–14, 2002.

[32]   Geiger PC, Cody MJ, Macken RL, Bayrd ME, and Sieck GC. Effect of unilateral denervation on maximum specific force in rat diaphragm muscle fibers. *J Appl Physiol* 90: pp. 1196–204, 2001.

[33]   Geiger PC, Cody MJ, Macken RL, and Sieck GC. Maximum specific force depends on myosin heavy chain content in rat diaphragm muscle fibers. *J Appl Physiol* 89: pp. 695–703, 2000.

[34]   Geiger PC, Cody MJ, and Sieck GC. Force-calcium relationship depends on myosin heavy chain and troponin isoforms in rat diaphragm muscle fibers. *J Appl Physiol* 87: pp. 1894–900, 1999.

[35]   Gosselin LE, Martinez DA, Vailas AC, and Sieck GC. Passive length-force properties of senescent diaphragm: Relationship with collagen characteristics. *J Appl Physiol* 76: pp. 2680–5, 1994.

[36]   Greer JJ, Allan DW, Martin-Caraballo M, and Lemke RP. An overview of phrenic nerve and diaphragm muscle development in the perinatal rat. *J Appl Physiol* 86: pp. 779–86, 1999.

[37]   Gregory SA. Evaluation and management of respiratory muscle dysfunction in ALS. *NeuroRehabilitation* 22: pp. 435–43, 2007.

[38]   Gunst SJ, and Tang DD. The contractile apparatus and mechanical properties of airway smooth muscle. *Eur Respir J* 15: pp. 600–16, 2000.

[39]   Gunst SJ, Tang DD, and Opazo Saez A. Cytoskeletal remodeling of the airway smooth muscle cell: a mechanism for adaptation to mechanical forces in the lung. *Respir Physiol Neurobiol* 137: pp. 151–68, 2003.

[40]   Gunst SJ, and Zhang W. Actin cytoskeletal dynamics in smooth muscle: a new paradigm for the regulation of smooth muscle contraction. *Am J Physiol Cell Physiol* 295: pp. C576–87, 2008.

[41]   Guth K, and Wojciechowski R. Perfusion cuvette for the simultaneous measurement of mechanical, optical and energetic parameters of skinned muscle fibres. *Pflugers Arch* 407: pp. 552–7, 1986.

[42]   Han YS, Geiger PC, Cody MJ, Macken RL, and Sieck GC. ATP consumption rate per cross bridge depends on myosin heavy chain isoform. *J Appl Physiol* 94: pp. 2188–96, 2003.

[43]   Han YS, Proctor DN, Geiger PC, and Sieck GC. Reserve capacity of ATP consumption during isometric contraction in human skeletal muscle fibers. *J Appl Physiol* 90: pp. 657–64, 2001.

[44]   Hebb DO. *The Organization of Behavior*. New York: John Wiley & Sons, 1949.

[45]   Henneman E, Somjen G, and Carpenter DO. Functional significance of cell size in spinal motoneurons. *J Neurophysiol* 28: pp. 560–80, 1965.

[46]   Hopkins SC, Sabido-David C, van der Heide UA, Ferguson RE, Brandmeier BD, Dale RE, Kendrick-Jones J, Corrie JE, Trentham DR, Irving M, and Goldman YE. Orientation changes of the myosin light chain during filament sliding in active and rigor muscle. *J Mol Biol* 318: 1275–1291, 2002.

[47]   Hudson AL, Butler JE, Gandevia SC, and De Troyer A. Interplay between the inspiratory and postural functions of the human parasternal intercostal muscles. *J Neurophysiol* 103: pp. 1622–9, 2010.

[48]   Huxley AF. Muscle structure and theories of contraction. *Prog Biophysics Biophys Chem* 7: pp. 255–318, 1957.

[49]   Iscoe S, Dankoff J, Migicovsky R, and Polosa C. Recruitment and discharge frequency of phrenic motoneurones during inspiration. *Respir Physiol* 26: pp. 113–28, 1976.

[50]   Jodkowski JS, Viana F, Dick TE, and Berger AJ. Electrical properties of phrenic motoneurons in the cat: correlation with inspiratory drive. *J Neurophysiol* 58: pp. 105–24, 1987.

[51]   Jodkowski JS, Viana F, Dick TE, and Berger AJ. Repetitive firing properties of phrenic motoneurons in the cat. *J Neurophysiol* 60: pp. 687–702, 1988.

[52]   Johnson BD, and Sieck GC. Differential susceptibility of diaphragm muscle fibers to neuromuscular transmission failure. *J Appl Physiol* 75: pp. 341–8, 1993.

[53]   Johnson BD, Wilson LE, Zhan WZ, Watchko JF, Daood MJ, and Sieck GC. Contractile properties of the developing diaphragm correlate with myosin heavy chain phenotype. *J Appl Physiol* 77: pp. 481–7, 1994.

[54]   Jones KA, Perkins WJ, Lorenz RR, Prakash YS, Sieck GC, and Warner DO. F-actin stabilization increases tension cost during contraction of permeabilized airway smooth muscle in dogs. *J Physiol (Lond)* 519(Pt 2): pp. 527–38, 1999.

[55]   Kannan MS, Prakash YS, Brenner T, Mickelson JR, and Sieck GC. Role of ryanodine receptor channels in Ca^{2+} oscillations of porcine tracheal smooth muscle. *Am J Physiol (Lung Cell Mol Physiol)* 272: pp. L659–64, 1997.

[56]   Katagiri M, Young RN, Platt RS, Kieser TM, and Easton PA. Respiratory muscle compensation for unilateral or bilateral hemidiaphragm paralysis in awake canines. *J Appl Physiol* 77: pp. 1972–82, 1994.

[57]   Kawai M, and Brandt PW. Two rigor states in skinned crayfish single muscle fibers. *J Gen Physiol* 68: pp. 267–80, 1976.

[58]   Kinkead R, Zhan WZ, Prakash YS, Bach KB, Sieck GC, and Mitchell GS. Cervical dorsal rhizotomy enhances serotonergic innervation of phrenic motoneurons and serotonin-dependent long-term facilitation of respiratory motor output in rats. *J Neurosci* 18: pp. 8436–43, 1998.

[59]   Kong FJ, and Berger AJ. Firing properties and hypercapnic responses of single phrenic motor axons in the rat. *J Appl Physiol* 61: pp. 1999–2004, 1986.

[60]   Kuei JH, Shadmehr R, and Sieck GC. Relative contribution of neurotransmission failure to diaphragm fatigue. *J Appl Physiol* 68: pp. 174–80, 1990.

[61]   Legrand A, Cappello M, and De Troyer A. Response of the inspiratory intercostal [correction of intercoastal] muscles to increased inertial loads. *Respir Physiol* 102: pp. 17–27, 1995.

[62]   Leiter JC, Knuth SL, and Bartlett DJ. The effect of sleep deprivation on activity of the genioglossus muscle. *Am Rev Respir Dis* 132: pp. 1242–5, 1985.

[63]   Levine S, Nguyen T, Kaiser LR, Rubinstein NA, Maislin G, Gregory C, Rome LC, Dudley GA, Sieck GC, and Shrager JB. Human diaphragm remodeling associated with chronic obstructive pulmonary disease: clinical implications. *Am J Respir Crit Care Med* 168: pp. 706–13, 2003.

[64]   Lewis MI, and Sieck GC. Effect of acute nutritional deprivation on diaphragm structure and function. *J Appl Physiol* 68: pp. 1938–44, 1990.

[65]   Lewis MI, Zhan WZ, and Sieck GC. Adaptations of the diaphragm in emphysema. *J Appl Physiol* 72: pp. 934–43, 1992.

[66]   Liddell EGT, and Sherrington CS. Recruitment and some other factors of reflex inhibition. *Proc Roy Soc Lond (Biol)* 97: pp. 488–518, 1925.

[67]   Mantilla CB, Seven YB, Zhan WZ, and Sieck GC. Diaphragm motor unit recruitment in rats. *Respir Physiol Neurobiol* 173: pp. 101–6, 2010.

[68]   Mantilla CB, and Sieck GC. Invited Review: Mechanisms underlying motor unit plasticity in the respiratory system. *J Appl Physiol* 94: pp. 1230–41, 2003.

[69]   Mantilla CB, and Sieck GC. Neuromuscular adaptations to respiratory muscle inactivity. *Respir Physiol Neurobiol* 169: pp. 133–40, 2009.

[70]   Mantilla CB, Zhan WZ, and Sieck GC. Neurotrophins improve neuromuscular transmission in the adult rat diaphragm. *Muscle Nerve* 29: pp. 381–6, 2004.

[71]   Miyata H, Zhan WZ, Prakash YS, and Sieck GC. Myoneural interactions affect diaphragm muscle adaptations to inactivity. *J Appl Physiol* 79: pp. 1640–9, 1995.

[72] Ottenheijm CA, Heunks LM, Sieck GC, Zhan WZ, Jansen SM, Degens H, de Boo T, and Dekhuijzen PN. Diaphragm dysfunction in chronic obstructive pulmonary disease. *Am J Respir Crit Care Med* 172: pp. 200–5, 2005.

[73] Pabelick CM, Sieck GC, and Prakash YS. Significance of spatial and temporal heterogeneity of calcium transients in smooth muscle. *J Appl Physiol* 91: pp. 488–96, 2001.

[74] Peter JB, Barnard RJ, Edgerton VR, Gillespie CA, and Stempel KE. Metabolic profiles of three fiber types of skeletal muscle in guinea pigs and rabbits. *Biochemistry* 11: pp. 2627–33, 1972.

[75] Pinto S, and de Carvalho M. Motor responses of the sternocleidomastoid muscle in patients with amyotrophic lateral sclerosis. *Muscle Nerve* 38: pp. 1312–7, 2008.

[76] Prakash YS, Gosselin LE, Zhan WZ, and Sieck GC. Alterations of diaphragm neuromuscular junctions with hypothyroidism. *J Appl Physiol* 81: pp. 1240–8, 1996.

[77] Prakash YS, Kannan MS, and Sieck GC. Regulation of intracellular calcium oscillations in porcine tracheal smooth muscle cells. *Am J Physiol (Cell Physiol)* 272: pp. C966–75, 1997.

[78] Prakash YS, Kannan MS, Walseth TF, and Sieck GC. Role of cyclic ADP-ribose in the regulation of $[Ca^{2+}]_i$ in porcine tracheal smooth muscle. *Am J Physiol* 274: pp. C1653–60, 1998.

[79] Prakash YS, Mantilla CB, Zhan WZ, Smithson KG, and Sieck GC. Phrenic motoneuron morphology during rapid diaphragm muscle growth. *J Appl Physiol* 89: pp. 563–72, 2000.

[80] Prakash YS, Miller SM, Huang M, and Sieck GC. Morphology of diaphragm neuromuscular junctions on different fibre types. *J Neurocytol* 25: pp. 88–100, 1996.

[81] Prakash YS, Miyata H, Zhan WZ, and Sieck GC. Inactivity-induced remodeling of neuromuscular junctions in rat diaphragmatic muscle. *Muscle Nerve* 22: pp. 307–19, 1999.

[82] Prakash YS, Pabelick CM, Kannan MS, and Sieck GC. Spatial and temporal aspects of ACh-induced [Ca2+]i oscillations in porcine tracheal smooth muscle. *Cell Calcium* 27: pp. 153–62, 2000.

[83] Prakash YS, and Sieck GC. Age-related remodeling of neuromuscular junctions on type-identified diaphragm fibers. *Muscle Nerve* 21: pp. 887–95, 1998.

[84] Raper AJ, Thompson WT, Jr., Shapiro W, and Patterson JL, Jr. Scalene and sternomastoid muscle function. *J Appl Physiol* 21: pp. 497–502, 1966.

[85] Rayment I. The structural basis of the myosin ATPase activity. *J Biol Chem* 271: pp. 15850–3, 1996.

[86] Rhoades RA, and Bell DR. *Medical Physiology: Principles for Clinical Medicine*. Baltimore: Wolters Kluwer, 2009.

[87] Schiaffino S, Gorza L, Sartore S, Saggin L, Ausoni S, Vianello M, Gundersen K, and Lomo T. Three myosin heavy chain isoforms in type 2 skeletal muscle fibres. *J Muscle Res Cell Motil* 10: pp. 197–205, 1989.

[88]   Sieck GC. Diaphragm motor units and their response to altered use. *Sem Respir Med* 12: pp. 258–69, 1991.

[89]   Sieck GC. Diaphragm muscle: structural and functional organization. *Clin Chest Med* 9: pp. 195–210, 1988.

[90]   Sieck GC. Physiological effects of diaphragm muscle denervation and disuse. *Clin Chest Med* 15: pp. 641–59, 1994.

[91]   Sieck GC, and Fournier M. Changes in diaphragm motor unit EMG during fatigue. *J Appl Physiol* 68: pp. 1917–26, 1990.

[92]   Sieck GC, and Fournier M. Developmental aspects of diaphragm muscle cells: Structural and functional organization. In: *Developmental Neurobiology of Breathing*, edited by Haddad GG, and Farber JP. New York: Marcel Dekker, 1991, pp. 375–428.

[93]   Sieck GC, and Fournier M. Diaphragm motor unit recruitment during ventilatory and nonventilatory behaviors. *J Appl Physiol* 66: pp. 2539–45, 1989.

[94]   Sieck GC, and Fournier M. Metabolic profile of muscle fibers in the fetal cat diaphragm. In: *Sudden Infant Death Syndrome, Risk Factors, and Basic Mechanisms*, edited by Harper RM, and Hoffman HJ. New York: PMA Publishing Corporation, 1988, pp. 361–378.

[95]   Sieck GC, Fournier M, and Belman MJ. Physiological properties of motor units in the diaphragm. In: *Neurogenesis of Central Respiratory Rhythm*, edited by Bianchi AL, and Denavit-Saubie M. Hingham, MA: MTP, 1985, pp. 227–9.

[96]   Sieck GC, Fournier M, and Enad JG. Fiber type composition of muscle units in the cat diaphragm. *Neurosci Lett* 97: pp. 29–34, 1989.

[97]   Sieck GC, Fournier M, Prakash YS, and Blanco CE. Myosin phenotype and SDH enzyme variability among motor unit fibers. *J Appl Physiol* 80: pp. 2179–89, 1996.

[98]   Sieck GC, Han YS, Prakash YS, and Jones KA. Cross-bridge cycling kinetics, actomyosin ATPase activity and myosin heavy chain isoforms in skeletal and smooth respiratory muscles. *Comp Biochem Physiol* 119: pp. 435–50, 1998.

[99]   Sieck GC, Kannan MS, and Prakash YS. Heterogeneity in dynamic regulation of intracellular calcium in airway smooth muscle cells. *Can J Physiol Pharmacol* 75: pp. 878–88, 1997.

[100]   Sieck GC, Lewis MI, and Blanco CE. Effects of undernutrition on diaphragm fiber size, SDH activity, and fatigue resistance. *J Appl Physiol* 66: pp. 2196–205, 1989.

[101]   Sieck GC, and Mantilla CB. Effect of mechanical ventilation on the diaphragm. *N Engl J Med* 358: pp. 1392–4, 2008.

[102]   Sieck GC, and Prakash YS. Cross bridge kinetics in respiratory muscles. *Eur Respir J* 10: pp. 2147–58, 1997.

[103]   Sieck GC, and Prakash YS. Fatigue at the neuromuscular junction. Branch point vs. presynaptic vs. postsynaptic mechanisms. *Adv Exp Med Biol* 384: pp. 83–100, 1995.

[104]   Sieck GC, and Prakash YS. Morphological adaptations of neuromuscular junctions depend on fiber type. *Can J Appl Physiol* 22: pp. 197–230, 1997.

[105]   Sieck GC, and Regnier M. Invited Review: Plasticity and energetic demands of contraction in skeletal and cardiac muscle. *J Appl Physiol* 90: pp. 1158–64, 2001.

[106]   Sieck GC, Roy RR, Powell P, Blanco C, Edgerton VR, and Harper RM. Muscle fiber type distribution and architecture of the cat diaphragm. *J Appl Physiol* 55: pp. 1386–92, 1983.

[107]   Sieck GC, Trelease RB, and Harper RM. Sleep influences on diaphragmatic motor unit discharge. *Exp Neurol* 85: pp. 316–35, 1984.

[108]   Sieck GC, Van Balkom RH, Prakash YS, Zhan WZ, and Dekhuijzen PN. Corticosteroid effects on diaphragm neuromuscular junctions. *J Appl Physiol* 86: pp. 114–22, 1999.

[109]   Sieck GC, Zhan WZ, Prakash YS, Daood MJ, and Watchko JF. SDH, and actomyosin ATPase activities of different fiber types in rat diaphragm muscle. *J Appl Physiol* 79: pp. 1629–39, 1995.

[110]   Trelease RB, Sieck GC, and Harper RM. A new technique for acute and chronic recording of crural diaphragm EMG in cats. *Electroencephalogr Clin Neurophysiol* 53: pp. 459–62, 1982.

[111]   van Lunteren E, Martin RJ, Haxhiu MA, and Carlo WA. Diaphragm, genioglossus, and triangularis sterni responses to poikilocapnic hypoxia *J Appl Physiol* 67: pp. 2303–10, 1989.

[112]   Watchko JF, and Sieck GC. Respiratory muscle fatigue resistance relates to myosin phenotype and SDH activity during development. *J Appl Physiol* 75: pp. 1341–7, 1993.

[113]   Zengel JE, Reid SA, Sypert GW, and Munson JB. Membrane electrical properties and prediction of motor-unit type of medial gastrocnemius motoneurons in the cat. *J Neurophysiol* 53(5): pp. 1323–44, 1985.

[114]   Zhan WZ, Miyata H, Prakash YS, and Sieck GC. Metabolic and phenotypic adaptations of diaphragm muscle fibers with inactivation. *J Appl Physiol* 82: pp. 1145–53, 1997.

[115]   Zhan WZ, and Sieck GC. Adaptations of diaphragm and medial gastrocnemius muscles to inactivity. *J Appl Physiol* 72: pp. 1445–53, 1992.

Author Biographies

Gary C. Sieck, Ph.D., is the Vernon F. and Earline D. Dale endowed Professor & Chair of the Department of Physiology & Biomedical Engineering at the Mayo Clinic. He is also Dean for Research Academic Affairs. He served as President of the American Physiological Society and President of the Association of Chairs of Departments of Physiology. He is also a Fellow of the American Institute of Medical and Biological Engineering. In the past, he served as editor-in-chief of the *Journal of Applied Physiology*, and he is currently the editor-in-chief of *Physiology*. NIH has continuously funded his research for more than 30 years, which focuses on the neural control of respiratory muscles. He has published more than 320 peer-reviewed papers, and he has trained more than 80 graduate students and postdoctoral fellows.

Heather M. Gransee, Ph.D., is in the beginning stages of her research career. She completed her Ph.D. in Biomedical Engineering at the Mayo Clinic in 2009 and is currently a postdoctoral fellow in the Department of Physiology & Biomedical Engineering. Her research focuses on neuromuscular adaptations of the diaphragm muscle. She is supported by a training grant from the NIH (T32 HL105355).